fingerprints...

THREE
FINGERED
PRESS
SAN DIEGO, CA

fingerprints...

a coffeehouse reader

Lew Decker

Published by Three-Fingered Press
www.facebook.com/ThreeFingeredPress

Available at
www.amazon.com

LIBRARY OF CONGRESS CATALOGING-IN-PUBLICATION DATA
Decker, Lew
fingerprints... a coffeehouse reader / Lew Decker
p. cm.
2009932444

ISBN 978-0-9840971-0-4

Printed in the United States of America on recycled paper

2 4 6 8 9 7 5 3

SECOND EDITION

Portions of this book have been published in *Cruising World*, a magazine for sailing enthusiasts, and in *CQ*, a journal for the amateur radio community.

Book design by Casey Clemens Decker
Text set in Palatino Linotype

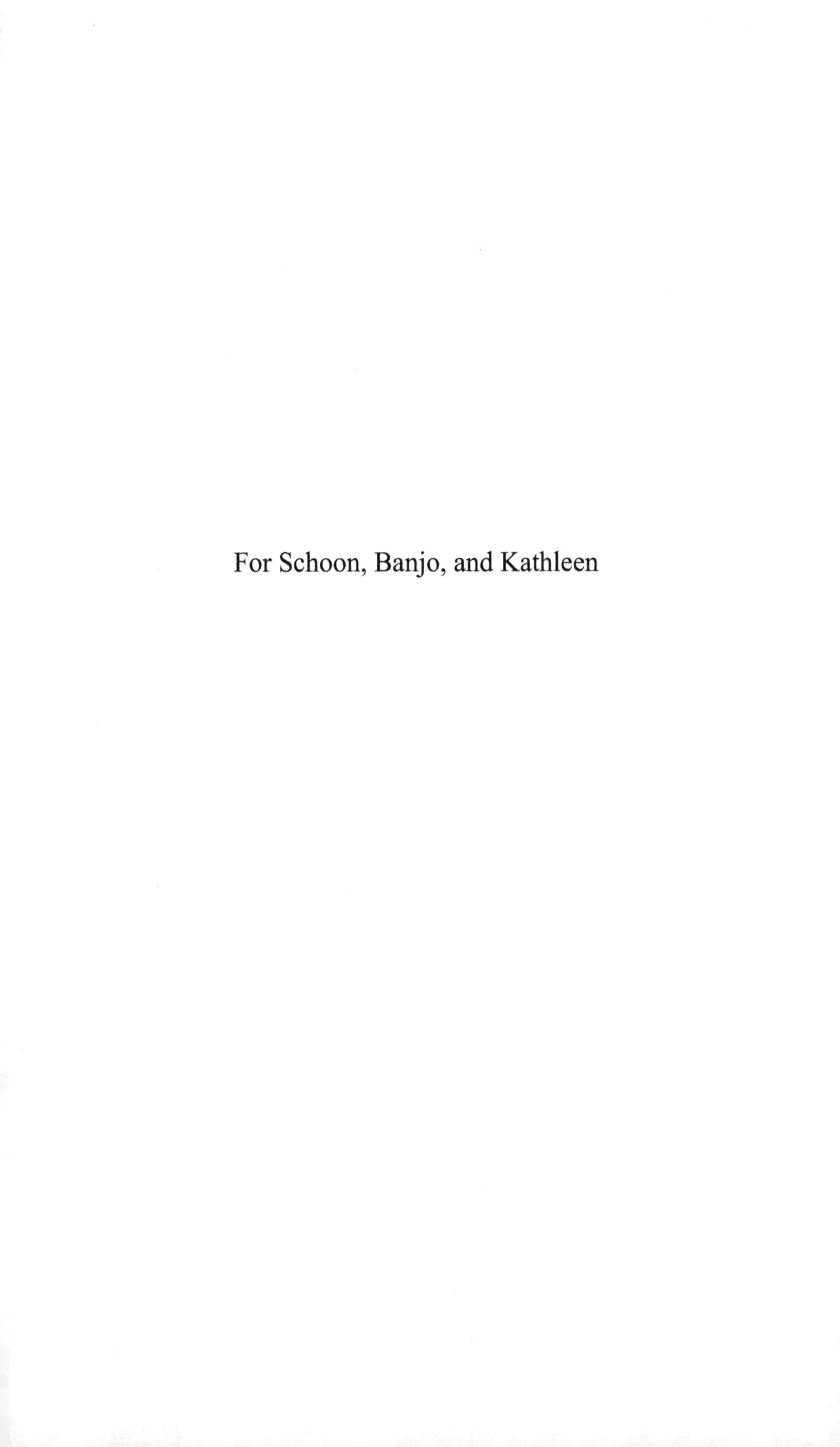

For Schoon, Banjo, and Kathleen

CONTENTS

author's note... XI

prologue... XIII

CHAPTER ONE
beyond the bridge... 1
CHAPTER TWO
big-hearted lady... 7
CHAPTER THREE
plaintive click... 11
CHAPTER FOUR
hiss and static... 14
CHAPTER FIVE
ghosting south... 19
CHAPTER SIX
little fires... 24
CHAPTER SEVEN
a regular customer... 29
CHAPTER EIGHT
barrier beach... 33
CHAPTER NINE
Tamarit... 39
CHAPTER TEN
drops in free fall... 42
CHAPTER ELEVEN
Joanie's note... 45
CHAPTER TWELVE
shore break... 48
CHAPTER THIRTEEN
Sun Flower...Two words... 52
CHAPTER FOURTEEN
no cure... 56
CHAPTER FIFTEEN
blue neon... 60

CHAPTER SIXTEEN
into the shadows… 64
CHAPTER SEVENTEEN
Jamaica then, my friend… 68
CHAPTER EIGHTEEN
the Zen of his own coffee… 71
CHAPTER NINETEEN
perfect circles… 75
CHAPTER TWENTY
trip into the night… 78
CHAPTER TWENTY-ONE
Sandy… 85
CHAPTER TWENTY-TWO
free in the wind… 87
CHAPTER TWENTY-THREE
slap and hiss… 92
CHAPTER TWENTY-FOUR
a bottomless pool… 99
CHAPTER TWENTY-FIVE
Willy Dog… 104
CHAPTER TWENTY-SIX
Big Max… 106
CHAPTER TWENTY-SEVEN
Ben Gunn… 111
CHAPTER TWENTY-EIGHT
field of flowers… 115
CHAPTER TWENTY-NINE
another kind of paradise… 121
CHAPTER THIRTY
Herbert… 129
CHAPTER THIRTY-ONE
staring from the rafters… 132
CHAPTER THIRTY-TWO
the fawn… 138
CHAPTER THIRTY-THREE
Sara Jane… 143

CHAPTER THIRTY-FOUR
another shot of Screech... 147
CHAPTER THIRTY-FIVE
feathers of spruce... 151
CHAPTER THIRTY-SIX
real palmiers... 157
CHAPTER THIRTY-SEVEN
ringing... 161
CHAPTER THIRTY-EIGHT
button faces... 164
CHAPTER THIRTY-NINE
typhoon... 168
CHAPTER FORTY
the binjo... 172
CHAPTER FORTY-ONE
Ranger Rick... 177
CHAPTER FORTY-TWO
crappy – all you can eat... 181
CHAPTER FORTY-THREE
heaven didn't work... 184
CHAPTER FORTY-FOUR
prance in silence... 189
CHAPTER FORTY-FIVE
the good Bustelo... 196
CHAPTER FORTY-SIX
the agonies of my father... 200
CHAPTER FORTY-SEVEN
sets on rewind... 203
CHAPTER FORTY-EIGHT
crystal mosaics, lightly smeared... 209

the cover... 215

the author... 217

AUTHOR'S NOTE

I took the liberty of changing the names in this book to avoid possible embarrassment for the people involved. A few of the place names have been changed for the same reason, but there are other issues at work. Because of the vagaries of my own memory, not to mention the passage of time, some of the elements in these stories may not have occurred in the sequence in which I have written them. I discovered after some soul searching that one of the stories is written backwards. Rearranging the sequence, however, would destroy the impact of what I tried to impart in the telling of it. I left it as it has been told. I hope you don't mind. Though I have been honest in writing my versions of these stories, it might be better if the reader considered this book to be a work of fiction. That way, no one's feelings will be hurt.

I never set out to write a book. I only set out to be a writer. My bride was impressed I would even try and, in the end, she decided I should put the stories into book form. It hasn't been easy to do. I spent a lot of time drinking coffee and trying to pare 160,000 words into something readable. What I took away from the exercise will stay with me for the rest of my life. For that I would like to thank all those kind people who poured my coffee and left me alone while I sat in the corner and tried to make sense of a lifetime spent wandering.

The kids at the Steaming Bean in Durango, Colorado, during the summer of '94 when I began this journey were a special bunch, greeting me each morning with a smile and a nod. The crew who ran Hot Java Café here in town were always there with a good sandwich and an Old Foglifter. I'm sorry their café had to close. The Juice n' Java in Pacific Grove is a quiet spot where I spent hours writing, rewriting, and watching the locals wander through. Over in Cave Creek, Arizona, I spent quite a few mornings in the Jackalope Café enjoying the high desert air and drinking coffee thick enough to obscure the bottom of the cup.

There were other coffee places (the Uphill Grind, the Pannikan, It's a Grind, Kensington Coffee Company, and the Metaphor Café) that provided safe havens, but most of my time was spent in Starbucks cafés scattered across the country. I never failed to find a happy lot behind the counter who were quick with the coffee and a good word for an old guy looking for a place to write. The crews at the Carmel Mountain Ranch Starbucks in San Diego have been exceptional over the years. I would like all of them to get a raise.

I'd also like to make space in these pages for my bride Kathleen, my son Casey, and for my daughter Marielle, who have humored me from the beginning. Without them, this book wouldn't exist. Without them, neither would I.

Thank you all for your kindness, your patience, and your good cheer. The journey has been grand.

PROLOGUE

On a winter night in Myrtle Beach during those bleak predawn hours, the outside air temperature dropped into the low thirties. A heavy rain that had fallen all night turned to snow. The cold woke me from a dreamless sleep, and I blinked into the darkness and shivered beneath the blankets while the clock ticked toward the alarm. I could see through the window from my bed but there was only the night sky outside, black as pitch-pine smoke. I curled up and waited for the screech of the alarm, hating that I had to get up to help my father.

When I stumbled into the kitchen, my dad was standing near the sink and staring at the snowflakes that floated in the dim light spilling from the back window. He shifted his weight and cleared his throat and leaned back with his hands on the counter. I looked up at him and then sat at the table where a bowl of soggy Cheerios waited. I ate all that I could before my father took the bowl away and rinsed it in the sink. He glanced through the window again and handed me a jacket.

Twenty-five miles to the north, up in Little River, a dozen stacks of newspapers waited on a corner for my dad. A stiff wind following the midnight front had driven the fresh snow into shallow drifts that blanketed the neighborhood. It buried the lone highway to the north. We left the house and groped our way along 38th Avenue toward King's Highway where my dad turned the car to the right and kept his foot softly on the gas pedal. The snow fell thick through the yellow headlight beams and made the station wagon feel like it wanted to slide on a sea of unsweetened butter. My dad drove slowly through the slop, convinced his papers would be at the intersection. When we got to Little River an hour later, though, there was nothing. He parked on the corner and turned the ignition off and sat for a moment. With the headlights darkened, all we could see was the new fallen snow glowing blue in the soft, ambient light.

My dad was afraid the newspaper truck would show up late and so we walked into the corner gas station to wait. The man who owned the station poured coffee for us. I took the cup from him even though the coffee was so strong and black and bitter I couldn't drink it. I stood quietly and held the burning cup and looked through the fingerprints on the window at the snow falling outside.

My father knew I didn't like coffee so he bought a bar of Hershey chocolate from the gas station man and handed it to me without saying a word. I bit into the chocolate and then sipped the bitter coffee and stood there with my dad. He held his cup with two hands while he stared through the windows at the cold and the snow and at the long, empty highway. I stared through the windows just like my dad and held the steaming cup near my face. The coffee with the chocolate was good then. We waited together and watched while the snow swirled in the wind and smothered the highway that faded into the gloom.

Our newspapers never did arrive that morning. The snow continued to fall and it took my father another hour to drive home. I was late for school again. When I finally settled into my seat in Mr. Rodman's second-period math class, I had to go to the bathroom but I was too afraid of him to ask permission to leave. I sat in the back of the room with my legs pinched together and waited for the bell. I could still taste the Hershey bar and the coffee while I sat there all pinched up with my face flushed from the heat in the classroom. I thought about my father and I thought about those long, barren highways. I thought about his face and how it was lined and drawn, and I felt again the winter wind laced with snow.

When I looked at the gray sky through the classroom window at the end of math period, the snowfall had stopped. I knew by the time we got out of school the snow on the ground would be melted. I sagged in my seat in the last row. Things would have been so much better if I could have stayed with my dad when the snow was still falling. The bitter coffee burned on its way down, but the chocolate made it sweet and I was content to stand in the station with my dad and look down the highway outside where there was nothing on the road except the wind and the snow. We stood together and watched through the windows and it was like we suddenly belonged there, just the two of us with our coffee, waiting for the newspaper truck and talking about the dreams we had, wondering about the future that always stayed hidden in the flurries just beyond the bend.

When I look back on my dad's time, I remember a lot of love and anger and confusion. I remember a lot of aimless wandering while he dragged my family onto the thin ice of financial insecurity, but the dreams that whistled away in the cold wind that morning stayed with me for a long time. Over the ensuing years I sat in a lot of cafés and stared out of a lot of dirty windows thinking about people like him and places I had been and things that just happened. It always seemed like I was back with my dad in that gas station, back with the bitter coffee and the sweet chocolate, back with those dreams that swirled away in the snow just beyond the fingerprints.

fingerprints...

CHAPTER ONE

beyond the bridge...

On the summer days beneath the drawbridge when the wind was down and the suffocating heat took hold, the catfish kids could be overwhelmed by the smell of decay. We could also be overwhelmed by the smell of the mud that looked and felt more like peat, and sometimes by the smell of the water itself, stained to the color of Italian roast by the tannins of rotting vegetation. In the shadows below the railroad trestle, the smells and all of that heat settled in like hothouse steam. It didn't matter much to us. We knew there were big flatheads lurking in the lee of the pilings steeped nearly black with creosote out near the channel. Scattered about in the shadows and the heat and in the wet, slippery mud, we waited together with our fishing lines sagging into the depths. The current swirled in and out and around the timbers, and the smell of the mud and decay drifted through like heavy marsh gas while we sweated out the catfish.

Sometimes the flatheads weren't enough to overcome the stench when something dead washed up in the shallows. That was too much, even for the older kids. We had to abandon the drawbridge and tramp along the banks to find a spot where the wind carried with it only the thick, sweet smell of the Carolina Lowcountry and reminded me of the hushed and beautiful cypress swamps simmering in the heat along the Waccamaw River over near Conway.

North of the drawbridge the Intracoastal Waterway was flanked on either side by muddy banks, high and steep, and the channel ran straight for nearly a mile. We sometimes hacked our way through the undergrowth to a spot on the bluffs where the air was clean and fresh. We had to cast our lines into the deep water far out in the center where the current was strong, but we got a good view of the river from up there. If the catfish weren't biting, there was sometimes enough Waterway traffic to keep us occupied.

Once in a while a tug wallowed down the river from around the bend, pushing a barge from behind and threading its way between the shallows. Pulsing with the exhaust from those big Detroit Diesel V-12's, the tug churned the bottom and drew torrents of mud and debris down the channel. We scrambled to get our lines reeled in before they were tangled and cut by the huge propellers. The captain usually waved at us as the tug chugged by, but there was always so much power and noise and water rushing through that when the tug disappeared downstream, the echoes it left behind hung briefly in the air and then faded into the heat and the heavy quiet of the Lowcountry.

Most of the workboats motored down the channel toward the south, throbbing like the engines of the Atlantic Coast line while they passed through the drawbridge, lumbering their way to Georgetown and Charleston farther down the Waterway. When the noise and the mud and the turbulence cleared away and the river returned to its penetrating silence, we cast our lines back into the water hoping for more catfish. It took a long time before we quit looking down the channel.

When the motor yachts came through, we had to stop our fishing to watch because none of us had ever seen such chrome and varnish and teak. It was hard for us to imagine having all that money. My brother and I sold popcorn at the Gloria Theater downtown while my older sisters worked as waitresses in the Seaside Café down beyond the Pavilion. The money we earned went to pay for school clothes and for family expenses because our parents were victims of the

Great Depression and of the war years that followed. They valued steady jobs and incomes, no matter how pitiful, and they valued these things even more than education itself. They certainly valued them more than freedom for little boys who only wanted to catch catfish. My mother used to tell us stories of walking twenty blocks to her job in San Francisco just to save a ten-cent cab fare and so some of the catfish kids worked out of guilt and some of us worked out of an innate sense of responsibility. Some of the kids didn't work at all and tried to survive on nothing.

The million-dollar motor yachts still rumbled by, bound for Miami and the Bahamas beyond or north to the Chesapeake. They plied the muddy Waterway in the midday sun, shining bright and polished with their spotless white hulls accented by cabin sides and cap rails of teak encased in layers of liquid plastic. Prominent East Coast yacht club pennants streamed from the mastheads and snapped at us in the wind like angry schoolteachers. Always there were deck crews in white uniforms moving about, looking starched and pressed and stiff, and girlfriends wearing tiny bikinis flopped around the decks in oily orgies of suntan lotion and shiny bronzed skin and Mai-Tai's in flowered glasses. All of the ragged kids back then, smelling of sour shrimp and slime and catfish, just stood on the muddy banks of the Waterway and watched.

Every now and then a big sailing yacht appeared from around the bend and sounded its horn. The bridge keeper usually yawned and looked away and the boat had to back down. Sometimes you could hear the captain yelling because a sailing yacht couldn't motor in reverse very well, but the little man up in the bridge office had his own agenda. I got a good look from the high banks at the scrubbed teak and at the lovely ladies sunning topless on the foredeck. I wanted to throw dead catfish over the rails when the boat came near just to startle the girls into sitting up, but

I only watched quietly as the yacht backed down and then quartered back and forth across the channel. I got a good view of the masts, with the rigging and the winches and all of those lines coiled about, and I loved those sailing yachts because it never seemed to me there could be anything so free as the wind except maybe your imagination. I stood with the catfish kids on the muddy banks of the Waterway aching to see those breasts, wondering how free you could be running off before the trades down in the Caribbean in a chrome and teak sailing yacht that would probably never leave the dock once it got to Miami.

Most of our fishing days were slow, though, with not much barge traffic to watch. It was a real event when one of the spit-shined yachts slipped into view from up the river. We worked the shrimp far out in the channels or in the softly whispering waters beneath the bridge, trying to keep cool, but always there was the slow-witted Southern afternoon that unfolded like a Peckinpah movie, bringing the kind of temperatures that made you feel like you were out on the two-lane dragging a ball and chain beneath the relentless Carolina heat. Sometimes even the catfish stayed in the mud. All we could do on those days was scan the river for barges or yachts, or the occasional ski boat that buzzed by, and dread the long hike home on blacktop that bled tar in the third-degree heat.

We hid in the shadows of the pines sometimes and watched our lines from a distance. If the drawbridge started clanging or the sound of the signal horns from river traffic drifted into the shade, we sprinted out to the edge of the banks to see what was coming. We were never disappointed since it took a big boat to get the drawbridge to open, but there was once a pair of heavy masts on a sailing yacht so wide it didn't look to us like it would fit between the pilings. When it came near, you could also see it was no ordinary yacht club queen. There were no ladies smothering breasts on the foredeck and no uniformed crew out polishing chrome and stainless. The only varnish you could see was on the cockpit coamings and on the immense teak wheel held steady by a bearded man in faded cut-offs who nodded and smiled after we waved. The masts of the schooner looked like telephone poles, and the hull was so massively built you had to know that she spent her life as a working schooner, not a yacht. From the ratlines and the baggywrinkle and the fishnet fenders you also knew there was a purpose here, that this was a real sailing ship meant for the trade winds that blow steadily across the Caribbean and for the heavy Atlantic swells that roll in all the way from Africa.

Just out of Georgetown thirty miles to the south, the Waterway opened to the sea. I tried to imagine that afternoon what it would be like to ride the foredeck of the schooner when it left the soft water behind and entered that zone where the ocean swells lifted the hull, where the wind filled the canvas and the spray off the bow stung your face. For the first time in my very young life I felt in touch with something other than theaters and popcorn and catfish and schoolwork, and something other than dreams about girls who never even knew my name. After the schooner had motored its way through and I was just a catfish kid again standing on the banks of the river, I stared for a long time at the empty channel thinking about all those islands down there and all that wind and all that water, blue as the neon sign on the theater marquee, and thinking about the power and the strength of the schooner that disappeared beyond the bridge, swallowed into the heat of the Lowcountry while it worked its way to the south.

Fishing the Waterway was never the same after that. I caught boatloads of catfish back then, mostly from beneath the drawbridge where the dense air settled thick and not so sweet in the shadows. A lot of tugs and barges and yachts passed through as well, but I was no longer content to watch them slide by from the side of the river. There was a longing somewhere inside triggered by the throbbing diesels of the tugs and by the pop-pop-pop of the chromed and polished yachts that plied the channel, triggered even by the whining of the outboard skiffs that disappeared so fast around the bend. The river traffic reminded me of the old, salt-bleached schooner that motored through on a hot afternoon heading for Winyah Bay and for the open sea beyond, heading for life itself hidden somewhere over the horizon. I didn't want to be left behind anymore, trapped beneath the indolent Carolina sun and standing alone at the edge of an endless ribbon of water so dark you only thought of coffee.

CHAPTER TWO

big-hearted lady...

A moment ago I looked through the window of the coffeehouse where I could see in the afternoon light a hot air balloon suspended in the haze that hangs over the hills. It was far away to the south and it had begun to descend and so I didn't see it for long, but I am a product of the Sixties. When I saw the balloon near the horizon, it was all about peace and love and tranquility and the Woodstock Nation even if there are rich people hanging in the wicker basket sipping from champagne flutes.

Hot air balloons seem like such works of art. The colors of the rip-stop nylon always remind me of Easter eggs and bubbles of soap and my very first carnival. When I saw the balloon in the distance, the red and blue and green fabric stood proud against the brown of the hills and the brown of the haze on the horizon and the olive drab of the valley oaks below, but then the balloon sank low in the sky and faded from view.

On one of those mornings when my brother and I waited for the catfish in the Intracoastal Waterway, we ran into another kid we knew from the neighborhood school. I had never seen him in the hallways wearing anything except old T-shirts and faded blue jeans and on the playground he stayed by himself and didn't say very much. I had the impression he was tough and streetwise and so I kept my distance when he was around. He asked that day if he could fish with us. Keith and I were afraid to say no and so he joined us on the high banks north of the drawbridge.

We didn't pose a very big threat to the catfish that day and after a few hours of heat and mosquitoes, this tough street kid wanted to know if we would like to go over to his house for lunch. When we agreed, he told us to wait. After he disappeared up the path through the trees that lined the Waterway, Keith and I sat in silence wondering what had happened. The kid returned in a few minutes and when he caught his breath, he told us he lived across the highway where the oak trees grew so big and that his mom would be ready for us in half an hour.

When we walked through the grove of oaks to the clearing beyond, it was like stepping into an old sepia print of a farmhouse cast adrift during the depression. Everything was brown and drab and colorless, very much like the hills that bake in the haze to the south of the coffeehouse. A grape stake fence that marked a small yard of dead grass had rotted to the ground where it lay like some neglected boardwalk. Most of the paint had worn away on the front of the house. The grain of the wood stood out, etched in relief like the siding had been sandblasted with time as relentlessly as it had been by the dust and the rain and the wind. Through the windows you could see some thin curtains swagged across the cloudy panes. They had been hanging so long there was no life to them. They sagged like the fence, worn and thin and faded to the color of dry concrete.

We stepped meekly inside to meet the mom. There was nothing cozy about the interior, just the brown planks of the floor swept clean and the shadow of a woman standing there who wore her eyes empty and her smile vacant and her dress drooping and shapeless like it hung from the rack at the Salvation Army thrift store. She stood in the kitchen wearing her exhausted face and her exhausted dress and looking at the three of us. She didn't say a word.

My brother and I sat down at the wooden table and

waited, not knowing what else to do. The air drifting through the house from the woman's kitchen had a warm and humid smell and it carried straight to my stomach like we were sitting at the counter in a café out on King's Highway. We watched the boiling pots on the stove and the steam swirled and the air hung over us heavy and moist while the kid's mom moved her weight gently around the room, so quick and efficient she reminded you of a sharecropper picking cotton before the afternoon rains.

When our lunch was ready, the woman came to the table with a large bowl of steamed white rice and another piled high with fresh butter beans. There was so little color in the house it seemed like the butter beans glowed in green neon and the rice in steaming white fluorescence, stark and luminous in the brown shadows of the kitchen. We were so young, though, and impatient, and we ate lunch so well and so fast that when we were finished, there was no more rice and there were no more butter beans and we were full and ready to get back to our fishing.

We said our thank-you's to this big-hearted lady who was so quiet and nice, and the streetwise kid turned to her as we left.

"Mom, that was good," he said. "What're we gonna' have for supper?"

This nice and quiet lady sagged against the counter and she looked away through the kitchen window.

"Son," she said softly. "We ain't havin' no supper."

I can't see the balloon through the window anymore. There are no more colors of the rainbow hanging in the haze and it's hard to imagine the wealthy people who were standing in the basket sipping champagne in the setting sun. Their money insulates them from the brown and from the drab and from the dark. For them there is only the carnival. For the rest of us, there is the brown haze out there sagging over the brown Southern California hills and the brown coffee that cools in my cup. The dark shadows remind me of an earlier time when the only color in a little boy's life was a plate of fluorescent rice piled high with butter beans, still glowing green in the dim reaches of a memory.

/) CHAPTER THREE /)

plaintive click...

The popcorn machine stood alone beneath the marquee of the Gloria Theater in Myrtle Beach. I was supposed to sell to the people going in to see the movies and to sell to the tourists who ambled by on the sidewalk, but there was a lot of down time in the popcorn business. I leaned against the machine and watched the traffic on the boulevard in front of the theater while the sounds of the cars that drove by mingled with the sounds of the German band organ that played in the amusement park across the street. To me they were just part of a circus parade that rambled through every night in the summer heat of the beach town. You could hear beyond the cars and beyond the band organ polkas the clacking of the roller coaster and the pops of the shooting gallery and the hum of the electric motors that powered the rides. Always there was the screaming and the laughing and the screaming again of the girls swirling happily in the sea of neon. From my side of the street, the carnival never seemed real.

Every night I stood near the popcorn machine and listened to those sounds while the eternal stream of cars that turned the corner down by the Bowery made its way up the hill past the theater and disappeared beyond the lights of the marquee. The parade never stopped until a dented and door-pranged Plymouth wheezed and coughed and sputtered to a standstill in the middle of the street. The driver of the car turned the ignition switch on and off and on again trying to restart the engine. You could hear the starter motor grinding strongly at first but the car wouldn't respond. The man kept turning the switch and the starter kept grinding slower and slower until finally there was nothing but the plaintive click of the solenoid and two spent headlights that stared into the night like a dead catfish.

Cars began to line up behind the stalled Plymouth. Someone honked a horn and someone else shouted out of a window and soon there were cars pulling into the other lane to go around and there was a lot of honking and shouting and swearing at the man who sat behind the wheel of the darkened hulk. A woman slouched in the passenger seat, staring through the windshield like the deadened headlights while the three kids who were in the back slumped forward with their faces down. I stood and watched and shifted my weight. Above the noise of the honking and shouting in the street, you could hear the mocking sounds of the carnival.

The man behind the wheel finally got out of his car and glanced at the lights of the amusement park and then shuffled over to the theater entrance where I heard the click of the nickel he placed on the ticket counter. He was a big man and when he stood before the window, I couldn't see around him into the ticket booth where Missy sat frightened and motionless. Neither of them said anything. I stared at the sweat stains on the man's back and at his shirt, worn thin and torn and too short to be tucked in, and at his pants that were too baggy in the crotch and shiny with grime. Missy didn't know what to do so she told him in her very small and quivering voice that he didn't have enough money to get into the theater. The man from the Plymouth still didn't say anything. He only slid his nickel off the counter and put his hands in his pockets and walked out from beneath the blinking lights of the marquee. He turned and walked up the street where he was swallowed into the night by the sounds of the honking of the cars and the shouting of the drivers and the screaming of the girls in the carnival.

The woman and the three kids he left behind in the middle of the street just sat there staring nowhere. Two policemen showed up. One of them tried to talk to the lady, but she only responded with more dead stares and so he turned and began to direct traffic around the hulk that was blocking the lane. The other policeman returned in a squad car with the lights flashing from the roof. He pulled in behind while the first took the wheel of the Plymouth and, slowly, they pushed the stalled car and the stalled woman and the stalled kids up the street into the heat and the darkness beyond the theater. I glanced over at Missy, not sure of what to say. She only sat there looking out of her window at the cars that were moving again. She rubbed her eyes and took a sip of her coffee and then she looked down at the book in her lap. I stood by the popcorn machine, wondering what had happened. Maybe the cars in the street had never stopped and maybe I was just dreaming. Maybe the only thing real was the carnival.

/) CHAPTER FOUR /)

hiss and static...

My job at the Gloria Theater paid me seven dollars each week and all the popcorn I could eat. I also got free admission to every Randolph Scott Western and John Wayne war movie that came to town. While my classmates tried out for the high school football team and flirted with the girls down at the Rexall, I sat in the air-conditioned darkness while the movies played endlessly the wailing sounds of dive bombers and the pops of small arms fire.

I fell in love with Mitzi Gaynor when *South Pacific* opened for a three-week run. The World War II tropical island fantasy marched across the screen while I sat through it eighteen or twenty times, swept away by coconut palms in glorious Technicolor and by stunning vistas of coral beaches in 70mm CinemaScope. Even the war gripped me from the screen. I understood the horrors involved but I still felt a sense of loss at the end of the film, like some tropical Brigadoon had settled briefly on an obscure island in the Pacific only to vanish just as quickly. After *South Pacific* closed, I thought about war and I thought about people like Emile de Becque and Lieutenant Cable, and I thought a lot about the short wave radio that let the theater audiences listen to a drama unfold off the screen. I was only a small time kid in a small town in the South, but the radio in the movie had a grip on me like the war itself. It made me think I could still be part of a Pacific island Brigadoon.

Sometime later during the school year I met an amateur radio operator whose call sign was scribbled on the blackboard in Mrs. Ferguson's algebra class. K4 "Young Dumb Operator" became my friend and mentor and a hero of sorts since he was an upper classman and a good student while I was just a freshman misfit. He had a 1944 Navy surplus Collins TCS-12 receiver and a modified Globe Scout transmitter set up in a space off his parents' garage. When I visited him that first time, I thought Emile de Becque's voice would crackle through the loudspeaker.

With my parents' permission I used some popcorn money to buy my own TCS-12 ($24.95 plus shipping from Arrow Radio in Chicago). K4YDO helped me put together a power supply out of one of my mother's aluminum bread pans. Our creation looked lethal, and it probably was, but the TCS sparked to life. The receiver was black, not Navy gray, but there were switches and dials and knobs built for abuse and a crackle-finished steel case cold to the touch. When I turned the radio on for the first time, the hiss and the static and the voices from far away flooded into my bedroom. I spent hours searching the airwaves lost once more in that *South Pacific* fantasy.

I lugged the big receiver to work one day so I could tune to the military weather frequencies for the updates on an approaching hurricane. The only customers we had in the theater that day were a handful of old men from off the street who had stumbled in to escape the wind and rain. I assembled the TCS in the lobby and turned up the volume. One by one the old men shuffled out of the theater and sat on the floor near the receiver with their eyes focused and their ears straining, listening through the static as the hurricane hunters radioed their reports. Outside of the theater the wind rattled the doors and blew the letters off the marquee, but inside the lobby a few surplus old men and a boy sat huddled over a radio that had been granted a second life. For a few hours that day, none of the street people needed anything to drink.

Eventually I took my Novice license exam and became KN4INW, operating on the 80-meter amateur band with the TCS receiver and a Heathkit DX-40 transmitter I built from a kit. Radio still seemed like magic to me and so I stayed in my World War II command post and exchanged Morse code signals with amateur radio operators all over the Southeast. I wasn't very good at fighting the interference at night when the 80-meter band was most active, so most of my operating was done after school and on Saturday afternoons when the band was quiet. One of my radio acquaintances told me to try the predawn darkness, so now and again I'd set my alarm for five o'clock in the morning and then drag myself out of bed to hit the airwaves in a fruitless chase for a station more than 500 miles away.

Out of the blackness one morning came a call from a K5 station in Tyler, Texas, the farthest contact I had ever made. We chatted for fifteen minutes or so while the band came alive. When our conversation ended, I copied a pair of stations from somewhere else out west. The signals weren't exactly weak, just distant, and they drifted into my bedroom wavering like the winds that send ripples across the Kansas wheat fields. I imagined an unpainted farmhouse and a split-rail fence and a sky that stretched beyond dreams, and an overalled farmer at his radio set pounding out his messages in the early morning quiet. I called repeatedly but to no avail, and in a short time the band closed up again. Disappointed, I dressed and trudged off to school. Like Texas and Kansas, though, everything seemed so far away, and I felt trapped in the classroom like the insects that exhausted their lives buzzing in the heat and shadows beneath the bridge over the Intracoastal Waterway.

When I finally understood enough electronic theory to pass the exam for a General license, the radio in my bedroom had lost its magic. There were no mysteries anymore, only a machine that spit static and another that spit radio waves into more static, and that's all there was. The end came when I let my Novice license expire without ever trying to upgrade even though my code speed was in the mid-twenties. I knew just enough theory to destroy my interest in radio.

A few years later my radio station was still intact in my bedroom, but a move to California was imminent. I put the gear up for sale. A buddy at the theater bought my beautiful TCS-12, and I threw in the bread-pan power supply with instructions on how to build a safer one. My DX-40 and the rest of the equipment I had accumulated went to a Canadian amateur who happened to be passing through town on vacation. I'm still mad at myself for letting that guy get away with my stuff for forty dollars, but I was intimidated by a man I didn't know who kept telling me he didn't think my transmitter worked anymore. He won, and I'm still not any good at business. Mitzi Gaynor and the command post faded away like the old men from the street who faded away after the hurricane.

Myrtle Beach happened a long time ago. A lot of water has flowed under the Highway 501 bridge since I bought that TCS-12. A good late-life crisis, though, can put distance between you and your past like nothing else. When these things happen, we scramble around and do dumb things trying to recapture those earlier times. We manage to get through the crisis and move on, but we usually accumulate a few toys and learn something about the kid who lingers in us all. I built an Italian racing bicycle from scratch because I could never afford one when I was young and riding every day. I still can't afford one, but I want one even more now than I did back then. It leans against a wall close to the wonderful seventies-vintage Atlas 350XL transceiver I bought some time ago.

Are these toys just one man's monuments to lost youth? Probably so, but I loved that old TCS-12. The local amateur radio store doesn't sell them. They have all kinds of digital shoe boxes that can squawk and pop in a dozen different languages, but whatever happened to cold steel and real switches and tubes that glow in the dark? Whatever happened to strength and character and charisma, and for that matter, Mitzi Gaynor? They are all there, I think, just beyond the knobs and meters of my Atlas.

A good cup of Italian roast goes well with a few minutes of dial spinning in the evening. When the recycled Atlas 350XL is switched on, an outside world comes to life again and I'm back to a time of hiss and static and old radios no one wants anymore. Sometimes when I hear a station with a southern call sign, I listen to the signals wondering whether the operator is someone I knew on the radio back then. I can concentrate so hard on the Morse code that my coffee gets cold. One of these days I'll no doubt sell the Atlas equipment and move to something more modern and complex, but for now I'm enjoying that command-post mentality I lost so long ago. You never know. There are a few of us out here who still believe in Brigadoon.

CHAPTER FIVE

ghosting south...

After Labor Day I could stand above the high tide marks where the sea oats grew thick and see forever along the beach that tapered unbroken for miles to the north. To the south the sand was deserted all the way to the center of town where the fishing pier and the Myrtle Beach Pavilion shimmered in the heat and in the heavy salt air. After the incredible crowds of summer had gone home, the solitude of the September shoreline made it seem as though the rush of the wind and the booming of the surf were long, heavy sighs of relief. The sounds made you want to flop in the dunes and listen to the shore break pop and hiss across the sand and to listen to the breeze drifting in off the sea that always brought with it the tingle of salt and the mystery of what lay beyond the horizon.

Surf fishing was good in the months after summer had ended. When you were a poor student trapped in the confines of a school where no one seemed to understand, all you wanted to do was leave your books in the locker at the last bell and race home just so you could get to the beach. I only had to grab a bucket and some frozen shrimp left over from the catfish expeditions on the Intracoastal Waterway and then walk the few hundred yards to the abandoned beach where I cast my line and sinker beyond the surf line at the foot of 38th Avenue.

I spent a lot of my time fishing alone and so I watched the sea change in its daily cycles. There were patterns that developed based on the ebb and flow of the tides and the strength of the wind and the time of the day, and so the sea, even in its infinite mood swings, became part of a friendly afternoon ritual as comfortable and as mysterious as the dark and slow-moving waters that swirled beneath the drawbridge out on Highway 501. I stood on the beach for hours and never once tired of the surf or the wind or the sound of the sea birds, and I never tired of searching the horizon for ship traffic.

Myrtle Beach lies about halfway between Cape Fear to the north and Cape Romain to the south and is near the center of Long Bay, a lazy hundred-mile curve of coastline that pushes inland away from the shipping lanes that pass farther out to sea. Surf fishermen rarely see the kind of traffic they might find during an afternoon on the Waterway. In the late fifties when there were no yacht clubs on that part of the coast, I never saw anything at sea except for the occasional shrimper. I kept watching, though, because even if all I ever saw were the shrimpers heading south out of Little River there was always a vague mystique about them, like the shrimp boats had some magical link to the sea and to the places down the coast that I didn't have. They always left me with a longing for the horizon and for what lay beyond, far out of sight for a kid with a bucket and a fishing rod and a classroom stuck in eternity.

When the bluefish started running in the fall, I moved my bucket and gear up to the foot of 43rd Avenue. Because of the sculpting of the beach sands, there was some deeper water just beyond the breakers where the schools swam closer to shore chasing baitfish and causing riots with the seagulls. The bluefish were always wary and kept a good distance from the beach. Even though I could cast far beyond the surf line, I had to wade into the combers to get close enough to reach them with my line. I caught a lot of bluefish out there, but it was an exhausting way to fish. When the frenzy of foam and diving birds calmed between the passing schools, I rested on the timbers of an old barkentine that lay buried in the sand.

During those moments of quiet, I sat on the bow of the hull that jutted from the line of dunes toward the surf. You could look far out to the horizon where the shrimpers ghosted through in the fading light. Those moments made me wonder about the timbers in the sand and how they got there and why the massive beams showed signs of fire. They made me wonder about the dreams of the men who sailed the old ship and whether they survived the fire and wreck. They also made me wonder about the sea itself and about the hurricanes, violent and unforgiving and absolute, that marched up the coast in the fall.

When the season of surf fishing ended, the Northers blew down with a vengeance. The driving rain came in wollops that made the beach look like the sound stage in an old black-and-white pirate movie. Most of the time I had to work at the Gloria Theater or stay inside and stare at homework, but one afternoon I read a brief column in the *Myrtle Beach Sun* about the exposed wreck up on 43rd Avenue. The article swept me up in the same kind of romance for the sea I felt after the bluefish had run through and the diving birds had flown away and there was nothing left but the beach and the thundering surf and the planks and frames of a story that lay buried in the sand.

The *Freeda A. Willey,* a 507-ton barkentine built in 1880 in Thomaston, Maine, spent most of her working life in the lumber trade sailing out of Pascagoula, Mississippi. Outbound for New York with a load of yellow pine, she was trapped offshore by "The Great Storm" of 1893 and abandoned at sea on August 28th after the crew cut the masts away when the hull became so waterlogged she was in danger of capsizing.

According to the *New York Maritime Register* of September 6th, 1893, a steamer captain reported seeing the abandoned barkentine twenty-five miles off Frying Pan Shoals. Even though there is no mention of fire, the ends of the timbers buried in the sand were blackened and charred. There was nothing left of the hull except the bottom planks and frames and the keel still sheathed in copper. One of the old-timers had said that at some point the galley stove overturned and caught fire and the barkentine burned to the waterline after the storm. The remains drifted ashore where they came to rest at the foot of 43rd Avenue.

With the endless cycle of fierce winter storms and occasional hurricanes in the summer and fall, the *Freeda A. Willey* could be buried under tons of sand for years or suddenly be exposed to the sun and wind and thousands of curious tourists when the sand was eroded by some change in that cycle. Most of the time when I lived there, a good part of the bow section was uncovered. When Hurricane Helene passed close offshore, the wind and tide removed the sand from the stern section as well. The misshapen hull lay open to the air and open to the neighborhood kids who swarmed over the hulk looking for pirate treasure. My brother and I spent a good part of our time crawling over the timbers with the other kids, all of us convinced there were thousands of gold doubloons buried somewhere just out of reach. There was such an aura of mystery around the wreck that when I got tired of digging around the sheathing and in the hull itself, sometimes I sat there by myself just wondering.

For the crew of the *Freeda A. Willey* there was no more romance, only the sky black as coal tar and wind that shrieked like dying men and waves that roared through like endless processions of serpents, screaming and raging and hissing in the night. In the end there were only the planks and frames and copper sheathing trapped on a deserted beach and a hollow feeling inside you for the sea and for what it could be. The timbers loomed naked from the high tide line like apparitions from some nautical Boot Hill, weathered by salt water and chafed by time until there was hardly anything left of the barkentine and nothing left of the Age of Sail. I suppose the timbers could have tempered with a small dose of reality how I felt about those shrimp boats ghosting south and about the sea itself that disappeared in secrecy over the horizon, but I was too young and naïve. All I ever felt was the mystery of places far away.

My family left South Carolina while I was still in high school. I didn't go back to Myrtle Beach for over twenty years. When I drove through again after all of that time, I had to stop and poke around the neighborhood on 38th Avenue, but I only stayed for a few minutes. I walked instead to the beach and turned north looking for the *Freeda A. Willey*, hoping to touch those timbers again and somehow reach back for that time when there wasn't much more to life than summer heat and sea water, warm as the womb, and bluefish racing beyond the surf line.

When I made my way to the foot of 43rd Avenue, though, there was nothing. Sand had piled so high it had reached the level of the sea oats growing at the top of the dunes. You couldn't tell there ever was a wreck. I walked around in the squeaky sand where I thought the timbers were hidden and I tried to calculate just how deep they were buried, but I gave up and looked out to sea like I had as a kid. I could see the empty horizon curving away, hiding all those places I had been in the years since. I turned back to glance at the sand one more time before I walked up to my car to go look for a coffee shop. I knew the old barkentine was there. It still held the mystery and the romance and the wonder for a ragged little surf fisherman who never tired of staring out to sea.

CHAPTER SIX

little fires...

It's clear today but there's a warm blanket of residual smoke that hangs low in the valleys. When I look to the mountains in the east the ridges are dark and green, but down along the lower slopes the line of haze is so thick it makes me think of water, and the peaks of the hills stand out like islands in the sun. There is only one brushfire burning now, southeast of us. The smoke drifts lazily through the valleys to flow beyond the coast and out to sea to the southwest where it spreads and thins and dies away. It will be days before the crews can clean up the hot spots that remain once the fire has been contained. The smoke will continue to drift in and out while the crews work and the arson team will poke around the ashes and the charred chaparral. Maybe they will piece together enough of the puzzle to arrest the kids who were playing with fireworks.

To the west of Myrtle Beach, the pine forest seemed as endless to us as the heat that never quit. We hiked for miles through the shadows of the trees all the way to the Intracoastal Waterway where we once found an old wooden shed, weathered and checked and abandoned. A construction crew had been clearing timber and bulldozing stumps, but they left the shed standing near the edge of the site to use as a storage container. The door of the shed was locked but through the broken window panes we could see several boxes stacked against the far wall nearly hidden in the dust and the cobwebs and the gloom. We thought the boxes were empty and so we took turns shooting through the windows like a sheriff's posse in a back-lot Western, but the BB's we shot didn't pass through to ricochet off the walls of the shed. They disappeared into the cardboard.

Our curiosity got the better of us. We climbed through the broken glass into the dust and the shadows to get a closer look. The boxes turned out to be heavy and I tipped one over to see the contents. The top broke open to spill a dozen foot-long rods, sticky with wax, across the floor. Ronnie Joe shouted with excitement because he thought we had found a case of highway flares, but the waxy sticks that tumbled across the floor weren't red, only an oily brown. We froze in our tracks until the dynamite stopped rolling around. Some minutes went by where we couldn't breathe. I took a very soft step toward the window, then another, and I climbed through the broken glass and took off running wide-eyed with my BB-gun pumping at my side. Keith and Billy and Ronnie Joe fell out of the window right behind me and ran hard to catch up. We stumbled over mounds of dirt and scrambled over fallen logs and crashed through the underbrush where the branches tore at us and the sweat and the dirt and the cuts made our arms sting, but we kept on running until Ronnie Joe shouted from behind for us to stop. I turned around to see my brother and Billy coming up, panting and covered with sweat. Ronnie Joe was doubled over, staring at the ground and spitting drool. We walked together after that and none of us had anything to say, but I had a dream that night and I could see myself standing in that shed where no one was moving and the sticks of dynamite were rolling across the floor. There was a massive and blinding white light and an ear-shattering report that was cut off sharp and I woke up in the dark, sweaty and confused and sick to my stomach.

The four of us talked about the dynamite for a long time after that. At first we made plans to go back to look for the field with the stumps and the bulldozers again and to steal some of the dynamite from the shed. We decided to make our own instead. Billy had a book from a chemistry set that explained how to make a crude form of gun powder so we went off to the drugstore to buy powdered charcoal and saltpeter. The clerk behind the counter wouldn't sell it to us. We did the next best thing and ran across the street to the fireworks stand on the corner where we pooled our money and bought rockets and mortars and a couple of Roman candles from an old man who didn't ask us any probing questions.

We spent the better part of the afternoon in Billy's garage tearing the paper and cardboard covers off the fireworks and then pouring the powder and the pellets into an aluminum pie pan. When we were finished, the pan was full and overflowing onto the concrete floor. We decided to make a huge bomb to set off on the beach, but we were curious and excited about all that powder and we kept talking about the best way to pack it and then we began to wonder if the powder would explode at all. The longer we talked the more impatient we got and then Billy lit a match just to see. There was a massive and blinding white light and a huge concussion and I was knocked backwards onto my back. Someone was screaming and then all of us were on our feet and scrambling out of the garage door because the ashen smoke was choking us. I couldn't see, only the white light of the explosion. I ran into Ronnie Joe who couldn't see, either, and we stumbled and fell into the grass in the front yard and rolled about, confused and scared and blinded. Billy kept shouting, "Holy shit...Holy shit," and Ronnie Joe kept asking if I was blinded, too, but in a moment my eyes cleared so that I could see a little out to the side. The smoke was still pouring from the garage because the slow burning pellets from the Roman candles hadn't yet gone out. My friends were sprawled on the lawn rubbing their faces and rolling about and then Keith started laughing. I couldn't help myself and we all rolled about laughing until our sides ached and the tears were streaming down our cheeks.

Hours later I still had that flash of light in the center of my eyes, but I sneaked over into Billy's garage and picked up the melted pie pan and tried to clean the powder burns off the concrete. The black spot looked like old crankcase oil and I didn't try very hard to remove it. I worried more about the odor and I was afraid of what Billy's father might do because now the garage smelled like the Fourth of July and there was nothing I could do. I couldn't understand why Billy lit the powder and I didn't understand why the garage hadn't burned down. I was more afraid one of my friends would tell and our fathers would beat us. Nothing ever happened.

It wasn't very long after that I lit a two-inch Salute and waited too long to throw it. When it exploded in my hand, I went into a state of shock and fell to my knees. My friends got scared and ran away and so I sat alone in the wet grass holding my hand and staring off into the darkness, barely conscious. Gradually the feeling returned to my fingers. If I had stuck my hand into a pot of boiling oil it couldn't have hurt any worse, but I was so afraid of what my father would do that I stayed in the grass and waited. Finally around midnight I gathered the courage to sneak home. When I opened our back door into the kitchen, my father stood there alone, leaning against the counter and drinking a bourbon and Coke. I couldn't look him in the eye. Tears welled up from somewhere and there was so much phlegm in my throat I couldn't speak. I stood quietly in the half-light of the kitchen doorway with my hand screaming and my body shaking in the late night heat.

On the counter stood the big white bowl my mother used for chicken bog. My father filled it with ice water and in a voice that was soft and deep, he asked if he could see my hand. When I walked over to him, he took me by the arm and placed my hand over the bowl. With his long fingers cupped like a ladle, he scooped the ice water and let it run down my forearm. The cold worked its way toward my hand and suddenly the pain wasn't there anymore, only a numbing cold that gradually felt more like warm bath water. I was sobbing then, and trembling, and the tears were hot on my cheeks. I laid my head against my father's chest and he held me and dripped the ice water slowly. My fear of him was all mixed up with my love for him and I couldn't stop the tears. It felt like that morning again in the snow when I stood with my dad in the gas station. The same warm feeling worked its way down when he turned away from the fingerprints on the window and smiled at me.

It has been more than fifty years since I stood in the kitchen with my dad and the ice water, and longer still since we stared through the fingerprints together. The brushfire to the south of the coffeehouse is nearly out now, but the other little fire, the one my father started, is still quietly burning.

CHAPTER SEVEN

a regular customer...

One night almost thirty years ago I sat alone in the Buena Vista Café in San Francisco with a line of empty Irish Coffee glasses in front of me. My parents knew the owners of the café when they were younger and I was thinking I should introduce myself to them again, but when another Irish Coffee showed up I sipped the drink and turned to watch the lights twinkling on Alcatraz while the whiskey burned its way south. My social graces were only adequate at best and worthless when I had been drinking, and so I sat in the booth by the window and kept to myself, wondering if George ever knew my dad had dabbled in moonshine.

I didn't know this either until I awoke from a deep sleep with a mouth so parched there was no saliva, a result of having spent six hours the night before eating popcorn while standing next to the machine at the theater in Myrtle Beach. All of the house lights were out when I sneaked into the kitchen so I opened the refrigerator door and let my eyes adjust to the soft light that flooded the linoleum. I didn't find any Kool-Aid but a one-gallon jug filled with apple cider stood at attention on the top shelf. Someone had put dried peaches in the cider and they hung suspended like golden prunes near the bottom of the jug. I didn't know why they were there. I lifted the jug and began swallowing cider by the gulp except that there was suddenly a burning sensation that scorched my throat. The gagging started soon after and it was all I could do to get the jug out of my mouth before I choked. I knew straight away that wasn't cider in the jug.

For about a year my dad worked at the local propane gas plant. I used to tag along with him when he made his weekly rounds in the delivery truck, a 1953 Chevy 3100 pickup with a huge propane tank bolted down where the bed used to be. He visited the trailer parks around town and filled the tanks for his customers and when the day was done we sometimes stopped at Causy's Flying A station downtown. I liked going with my dad on those occasions because there was usually a pot of chicken bog cooking on the propane burner in Mr. Causy's garage. There was also something brutal about the Ford Fairlane Crown Victoria parked in the far corner. It sat sulking by itself, hunkered down over its Goodyears like a Nascar stocker. I couldn't keep my eyes off of it.

A few days after the apple cider incident I asked my dad about the Ford parked in the Flying A garage. I was also suspicious about the jug of bootleg whiskey in the refrigerator. My dad smiled at me and told me that Causy had a little business on the side and that the big Ford was used to make deliveries. He wasn't going to say anything else but I didn't want to let the subject drop.

"What kind of business is that, Dad?"

"Well," he said. There was some hesitation where he took a breath and then he added, "Causy had a whiskey still a couple of years ago over on the Waccamaw, maybe ten miles below Conway. He used to fire the mash with pitchpine, but the Revenuers took to flying over the trees in an airplane. They could spot the pitchpine smoke from miles away. They didn't get Causy, but they broke up the still and put it out of business."

My father didn't want to say anything more. I sat in the passenger seat of the propane truck thinking about whiskey stills and thick, black smoke and wondering if the big Ford was ever used again to make deliveries after the Revenuers came through.

When we wheeled into the front of the Flying A a few nights later, I stepped out of the propane truck and stood next to the Crown Vic hiding in the service bay. I could hear it ticking and creaking while its engine cooled down. The right front tire was still warm to the touch. I sat on a folding chair and waited for my dad, but the Ford sulking in the corner was all I could think about. On the drive home I wanted to ask my dad again about the Ford and about Mr. Causy's other business but I looked through the window instead. When we stopped in our dirt driveway my dad pulled from behind the seat another jug of whiskey with the dried peaches floating on the bottom.

I gathered up a little nerve and said, "I thought that stuff was apple cider, Dad. I took a big swig the other night and nearly choked to death. What are the peaches for?"

"Causy puts them in to give the whiskey a little color and to soak up the tar. I don't know if it works. Did you get sick?"

"No, I guess not. I fell asleep. Is Mr. Causy still in business? Is that why the Ford was still hot?"

"Remember when I told you about the pitchpine smoke? I got to thinking about propane burners and how there isn't any smoke."

My father stopped in the driveway and shifted the jug of whiskey to his other hand. He reached over and in a rare show of affection he rubbed the top of my head.

"I sold him a pair of big propane burners and a 200-gallon tank. He's a regular customer, now."

I don't remember how many Irish Coffees I drank that night thirty years ago and I never introduced myself to the owners. If I had I would have been thinking about the pitchpine smoke and the propane burners. While my dad was struggling to survive in the South, George and his business partner were serving up what would prove to be a huge financial success. Irish Coffee at the Buena Vista Café became an institution in San Francisco. White lightning is a lot closer to gasoline than to apple cider, or to Irish whiskey for that matter, but it's just as big an institution in the South. It didn't make much difference to me that moonshine was illegal. I'm sure the owners of the Buena Vista Café were proud of their contribution to the local drinking culture. I've always been proud of my dad and his.

CHAPTER EIGHT

barrier beach...

There is a neon sign on the back wall of the café that flashes "Kensington" in cursive and "Coffee Company" in print. When the light comes on it brightens the counter and splashes its red and blue across my table. If I listen closely I can hear the sixty-cycle hum over the mandolin music on the sound system. There was a big neon sign over the theater where I worked as a kid that used the same colors as the Kensington sign. The sixty-cycle hum was so loud you could hear it up the street. I spent so much of my time there the sound of the hum and the lights of red and blue became part of my soul.

The people who ran the movie houses in town had been old vaudeville performers. To them the movie theater was still show business. They never talked about the old days, but you had to know there was too much travel and too many late nights and too many drunken crowds to please, and you also had to know they were cheated out of a lot of their money. They aged in the show circuit and became lined and dumpy. When the vaudeville curtains faded away and the stages fell silent, they wound up working the movie theaters in the beach towns. It was all the same show business to them except they drank unmercifully and the smell of bourbon in the office and on their breath was constant.

I worked for the vaudeville performers for over five years. I stood beside the popcorn machine each night and listened to the hum of the lights on the marquee that clicked on-and-off the letters in red and blue neon that spelled

"Gloria" in cursive and "Theater" in print. There were other sounds as well; popcorn spilling from the kettle in the machine, rich kids with Chevy 283's modified with headers and lakes pipes rapping down the hill in front of the theater, and the endless screaming in the carnival rides across the street in the amusement park. I never heard the little black kids, though, sitting on the curb with their backs to me who were not allowed to cross the street to play in the white man's carnival, or enter the white man's penny arcade, or even swim in the white man's ocean at the foot of the hill. They just sat on the curb, mute. Behind me there is the hum of the Kensington Coffee Company sign that drifts in and out with the music of the mandolins in the background. It sounds very much like the theater days again, hum and silence, hum and silence: the hum of the neon that cast the curb in shadows of red and blue, and the silence of the little black kids who were trapped beneath the smothering veil of white.

Somewhere in a box in a storage unit close to my house I have a movie program hidden away that came from the old Gloria Theater in Myrtle Beach. Printed across the bottom in bold type are the words BALCONY FOR COLORED PATRONS. I don't know why I've kept it all these years. Maybe it's just a reminder of a time when my view of the world was more innocent, a time when the only questions I ever asked were about the catfish in the Santee River or about the cedar waxwings that migrated through or about the spots and croakers that swam in the fall beneath the pilings of the 2nd Avenue fishing pier. I didn't know color. When my boss at the theater asked me if I would work the balcony as the usher and ticket boy, I climbed the stairs and sat near the aisle and collected tickets and helped people find their seats in the dark and watched the movies when no one came. Sometimes the cooks or the dishwashers from the Bowery down the street brought me foot-long hot dogs with mustard and relish and onions. I sneaked the guys in to see the movies for free. I never told the vaudeville performers about that.

A few years ago I returned to the South just to see how much had changed. We drove up King's Highway from Charleston and stopped for a look on the bridge just south of Georgetown. The Santee was one of the rivers of my childhood, much like the Waccamaw that emptied into Winyah Bay and the Intracoastal Waterway that snaked beneath the drawbridge on Highway 501.

I usually think of those rivers in terms of catfish and mosquitoes and water the color of espresso and the kind of heat and humidity that can drain away ambition and leave you sprawling and drenched in sweat that never stops, but it was cloudy the day we stopped near the bridge. The smoke from the pulp mill in Georgetown hung in the air and muted the colors of the water and the colors of the trees. When I looked up the river, it seemed to vanish in the haze like fading memories, watered down in so much distance and time the details became blurry and indistinct. There were no cloudy days like that when I was young, only sun and splashing and laughter and kids whispering in leaky wooden skiffs and the heat, always the damp heat, hanging low over the water.

To my relief, the Santee looked the same. There weren't any hotels where we crossed and there were no marinas and no boat traffic or blinking neon signs, only a wide slow-running river lined with trees hanging limp with Spanish moss, a river that turned gray in the haze and disappeared beyond a bend into the cloud cover and the smoke of the pulp mill. I tried to see the river again through the eyes of a twelve-year-old. Even in the haze and the gray of passing time, there were still images of backcountry swamps and cane poles bent double and hushpuppies and catfish and bowls of Mr. Causy's chicken bog.

Beyond the Santee and the Waccamaw Rivers, we drove north into Murrell's Inlet where you could see across the cordgrass the big homes that had been built on the barrier beach. Many of the houses out there are the summer cottages of the rich and when I first saw them when I was young, I felt like a street kid looking through the display window at Chapin's Department Store in downtown Myrtle Beach. There was never enough money for us back then. We had to work early on to survive and it set us apart from the people who had been born and raised in the South. Many of them came from old money. They led their lives as if they were landed gentry who had little time for the working class even though the old money had been gone for decades. My mother came from a Southern heritage, yet even she felt the barriers that separated us. We worked for wages and lived in rented houses and drove used cars and were never accepted into the community. Our lives played out in a separate world while the people who rattled around in the big houses hardly knew we were alive.

There was always the one barrier, though, I never understood. It made me realize later that, in some ways, my mother would forever be Southern and no matter how much I loved the South, I would never be anything more to the people than just another misplaced Yankee kid passing through.

The caretaker of the property next door to us lived in a tiny room off the garage in back separate from the main house. He kept to himself and most of the time we never knew he was around. He worked the lawns and tended the shrubs and drifted away in the evenings and returned late at night. He rarely spoke to us. If we called to him he would smile and wave in return and he would fix a bicycle if you asked or untangle the backlash from a fishing reel, but mainly he was just Major who worked in silence on the big yard next door.

One night my father came home late after drinking white lightning again in Mr. Causy's garage. He began to argue with my mother who was washing dishes in the kitchen when a soft knock came at the back door. My dad opened it and found Major standing alone in the dark. My father invited him in for a shot of moonshine and some Coca Cola, but Major declined the Coke and took his whiskey straight from a water glass with no ice. My mother was polite and stood quietly in the corner and watched. My father was very drunk and he kept asking Major if he would like another shot and Major kept taking the shots and slugging the whiskey back and he got so drunk so fast it scared my mother. He still didn't say much, but he couldn't stand straight any longer. When my father's whiskey was gone, Major thanked us quietly and then he asked if someone could drive him into town.

My father couldn't stand without leaning against the counter so my mother helped Major into the car. I hopped into the back to be with my mom because she was so nervous, and off we went into the heat and the darkness toward a part of town we had never seen. Inland and away from the sea, there was a segregated area where the people lived who worked as domestics in the motels and lodges on King's Highway and in the hotels that lined the beach. It was a quiet community of grocery stores and gas stations and schools and churches. Major wanted to be driven into the center of the town to a bar that hid in the shadows of a red and blue neon sign that said PRIVATE CLUB with the A-T-E burned out. My mother had trouble finding her way, but Major was a good and patient man and when we finally picked our way to the street where the neon club sign flashed into the night, Major turned to my mom and looked at her for a moment. My mother pressed against the car door.

In a voice so low you could barely hear, Major just said, "God bless you, Ma'am," and he opened his door and stepped from the car and staggered away toward the bar. My mother watched him go and then she turned and glanced at me in the back seat. I could see the fear in her eyes and I didn't understand the reason for it. She turned the car around and we sped away from Major and the blinking signs. I sat in the back and didn't say anything.

The summer heat still shimmers off the asphalt in the parking lot outside. When I leave the coffeehouse in a few minutes, I will have to re-enter my world and wade through the heat to get to my truck parked next to someone's Mercedes. The sun will remind me of the South again and the memories will be warm like the days I spent under the drawbridge when I had nothing but a cane pole and some patched Levi's and chicken bog to eat again when I got home while other people had huge houses on the beach to use in the summer and Cadillacs to drive on Ocean Boulevard. Even now I would love to have the option of owning some of those things, but in the South when I was first learning to ask questions, there were far more serious barriers than money.

CHAPTER NINE

Tamarit...

All of the heat and the dark water and the catfish and all of the cypress swamps that we got used to while living in Myrtle Beach gave way to what the Monterey Chamber of Commerce liked to call "nature's own air conditioning". My brother and I nearly froze to death. Where the beaches in South Carolina extend for mile after white sand mile, the coast of Monterey and Pacific Grove is broken up by jumbles of rock, cold and hard, and by short stretches of brown sand covered by rotting kelp buzzing with millions of flies attracted to the stench. Sometimes the water in South Carolina reached eighty-five degrees during the summer, but in Pacific Grove I don't know that it ever reached seventy. You couldn't dive without a wetsuit, even in August. The California coast was starkly beautiful, though, and it attracted the wealthy and famous who lived all over the Monterey Peninsula. When my family moved there in 1961, we were as out of place as we had been while living in the South.

You had to love the Pacific because it dominated the very soul of the Peninsula, but when Keith and I went down to climb on the rocks for the first time, we were stunned by the clarity of the water and shocked by its temperature. Our high school in Pacific Grove was crawling with surfers and when the winter storms offshore drove the surf to unusual heights, there was a considerable absentee rate. Neither my brother nor I ever got used to the drop in temperature. Keith was a lot tougher in that regard and he learned to surf and dive in the frigid waters. I could only remember those sultry mornings on the Waterway and the afternoons spent surf fishing when the eighty-degree air was cooler than the water. The South had sneaked up on me and so my dreams of sailing and the sea always included fantasies of heat and trade winds and water warm as a bathtub and miles of beaches where the white sand was so fine it squeaked beneath your feet. I grew to love the hard, brutal rocks of Pacific Grove, as well, and the short stretches of beach with the sound of heavy surf that thundered all the way to our house a mile and a half up the hill, but the Pacific remained distant and cold and ominous, like I was still a child trembling after a nightmare.

The wharf in Monterey was always filled with tourists looking for souvenirs and cotton candy. Out toward the end, though, where the day-boats waited for whale watchers and bottom fishermen, the crowds thinned out and I could lean by myself over the rail behind Angelo's Fish Grotto and look down on the deck of a seventy-two foot schooner. *Tamarit* never went anywhere, but when I remembered how it was to watch the shrimp boats on the horizon while standing on a lonely beach and how they made me long for places I had never seen, a schooner this close made me ache in the pit of my stomach.

I used to sneak down to the wharf just to see if anyone was aboard the *Tamarit,* hoping against all logic that the skipper would understand the ache I felt and invite me aboard. It never happened. I thought so much about the schooner behind Angelo's that it seemed like we sailed together all the way from Boothbay to Sidney, but I went away for a while and when I came back, the *Tamarit* was gone. In its place was another diesel-powered day-boat for the tourists to chase the gray whales or to fish the bottom for rock cod and sand dabs.

To this day when I return to Monterey, I can't visit the wharf without thinking of the *Tamarit* or where she might have gone. I sometimes stop at the coffee kiosk at the front of the wharf and then walk out to the end where I can lean on the rail and sip the Italian roast that steams in the cold northwest wind. The *Tamarit* is never there, but like the shrimpers that ghost along the South Carolina coast and disappear into the twilight, the schooner is still a link to all those places hidden beyond the fading horizon. I can lean on the railing and look across the harbor to the breakwater and the mouth of the bay while I hold the coffee cup with two hands and stare through the steam at the open sea in the distance. Sometimes there are big sails out there and I wait just to see, you know, if they belong to a seventy-two foot staysail schooner.

/) CHAPTER TEN /)

drops in free fall...

I drove through Pacific Grove this morning on my way to Juice n' Java down on Lighthouse Avenue. The fog was so thick it hung over the houses like Halloween lace, with the green of the trees looming stark against the gray. You get a feeling of self on mornings like this, maybe because the fog shuts out the world and there is only you and the trees and the houses close by fading in and out of the lace. All else is gray and nothingness and shadow. Memories form like the drops on the eaves and on the trees where they hang briefly and then free fall to the ground. More droplets form behind them, one after another, and you move slowly and think of the silliest things, one after another, because there is only you and the fog and the drops in free fall. Sometimes a morning chill creeps up on you then and it's so cold and gray and still outside it seems like there should be a foghorn somewhere in the distance.

My grandfather lived on 23rd Avenue in San Francisco in one of those classic row houses. I remember sitting in his little back yard and hearing someone play the organ. Every morning you could hear the music and it sounded to me like a church. I finally asked my grandfather why that guy kept playing the same two notes over and over, and he took me by the hand and we walked out into the alley. He pointed toward the sea lost in the gray and he told me there was no organ, only a foghorn to mark the point of land at the entrance to the bay. Now, whenever the fog rolls in from the sea and I am shut in, I always think of holding hands in the alley with my grandfather who was a beautiful man and of the foghorn out there in the gloom that played again and again the same two notes that made me think of Sundays.

There used to be a foghorn here in Pacific Grove, down the hill toward the beach in front of the lighthouse. When I lived up on Funston Avenue, you could hear the lonely church-organ sounds day and night because the fog on this part of the Central California coast is so dominant it becomes part of your life. In the years after I moved away, though, attitudes changed somehow and the locals got tired of the monotonous sounds. Maybe they even lost their sense of romance for the sea. There were some complaints and then the city council got involved and so the Coast Guard built a baffle on the inland side of the horn to muffle the noise. The baffle didn't work very well, though, and on foggy days you could still hear all over town the monotone notes like the foghorn in San Francisco that always sounded like a church organ.

The tiny building that housed the horn mechanism is still there, but with the advent of the Global Positioning System the Coast Guard decided the foghorn was useless and they shut it down permanently. For a long time you could see the baffles that didn't work hanging over the roof like a cormorant drying its wings, but now the baffles are gone and there is no longer any sound to the fog except the drops in free fall and the surf that booms along the rocks in front of the lighthouse.

You can walk from the coffeehouse in Pacific Grove south along Lighthouse Avenue until you go over a small hill where there is a cemetery and a golf course and a grove of Monterey cypress. There are small herds of coastal deer feeding among the headstones and along the fairways and sometimes in the ice plant that grows thick near the old foghorn. The streets are quiet early on and so the deer will drift about, shy and skittish, but if you're careful and move slowly they will stay while you walk down the hill toward the beach where the fog hangs soft through the trees. You can hear the roar of the surf there and the salt in the spray off the rocks mixes with the decay of the loose kelp that washes in with the tide. The smell in the air is cold and damp and heavy, like the smell of tarred hemp lying awash on the deck, and you can move quietly into the morning with the deer and the fog and the silence of the trees. If you listen closely again, you can hear all those droplets in free fall.

CHAPTER ELEVEN

Joanie's note...

The notes of a classical guitar are echoing through the empty Starbucks, like someone playing music in a stone courtyard somewhere around the corner. It's so quiet that if you shut your eyes you might expect to hear the dripping of a fountain or the footsteps of a stranger, or the voice in the distance of someone you haven't seen in years. Beyond the coffeehouse windows, a layer of fog has been drawn inland from the sea. The morning is gray and mellow and cool, and there are no sounds to the outside, only the guitar.

A friend of mine from a long time ago wrote a letter to Joan Baez when she first began singing professionally. In his note he asked if she thought her guitar might have a soul. She sent to him in response a handwritten letter that was gentle and warm and caring. She told him her guitar didn't have a soul, but that sometimes when she played, her guitar became an extension of herself. Through her guitar, she was able to be free. David kept the note pinned to the wall in his room and I stopped to read it each time I was there. David didn't live very long, though, and when he died, his father took the note down and packed it away in a trunk and carried the trunk to the basement. The note is probably still there.

Joan Baez kept a low profile as a part-time student at Monterey Peninsula College, the school I attended when I was nineteen. I didn't know her personally, but I would see her now and then in the library and always the kids in the school would ask her to sing. One day toward the end of the semester, Joan brought her guitar and during the lunch break she began to sing on the green in the center of the campus. Slowly and quietly the students of the college drifted onto the grass from the library and from the classrooms and soon there were several hundred of us. We sat together in silence and listened to Joanie's guitar and to her soul.

Sometime during the concert, two Marines in uniform showed up and stood at the back of the crowd to hear. One of the students saw the Marines and so he left the group and returned shortly with a large piece of white butcher paper on which he had scrawled the words MARINES KILL in black marker pen. The butcher paper began to circulate among the students. Each of them signed the sheet and passed it on to someone else. By the time the Marines became aware of what had been written on the paper, it was filled with signatures. I watched from the back of the crowd as the Marines exchanged glances and then they turned with their heads down and walked away.

For a time those Marines had been part of our world, part of a community that had been shaped and molded into one by the soft sounds of Joanie's guitar and by the soft sounds of her soul. I remember thinking how it was to sit on the green beneath the spring sun and to be part of all that is good about the human soul, that we could set differences aside and become one through the music of a lovely girl with a gift, but then it was gone. There was a war going on in Vietnam that no one wanted and the few moments of peace we had together on the green ended with an ugly reminder that there was no peace.

Joanie continued to sing that day, but after the Marines went away, there was no point to the banner and there was no longer any gentleness to her voice. You could hear the sharp edges of protest and I wondered about soul again and about guitars and about Joanie's note pinned to the wall in David's room.

Joanie's note...

When I left the school that day, spring seemed a long way off. There was something ominous about the soul Joanie had bared. I didn't understand the concept of war and the Marines who had listened so quietly from the edges of the green seemed only like my father who had been a career military man. Some of my friends were already in Southeast Asia and some had run away to Canada and my draft deferment was to be canceled in a few months. The confusion I had then was like the confusion I felt when Kennedy was shot. I only wanted to sit with my friends in the sun and listen to Joanie. I wanted to hear from her soul the words of poets and the music of peace, but in the end there was only a banner for Marines who kill and a gut wrenching war and a note packed away in a trunk.

CHAPTER TWELVE

shore break...

In early December of 1965, I drove a sputtering Morris Minor to my first Navy duty station below the 28th parallel on the South Texas coast. The temperature on the day I arrived hovered near ninety degrees while the humidity climbed even higher. Corpus Christi sweltered like August in the Carolina Lowcountry. I checked into the Supply barracks at the Naval Air Station during the hottest part of the afternoon and found a bunch of guys lolling about on the second deck of the building waiting for the sun to go down. The barracks weren't much cooler than the air outside.

I introduced myself to a swarm of sailors in white T-shirts and white boxer shorts who welcomed me into my temporary berth in the barracks. One of the guys looked very much like Mr. Peepers of television fame, but he drove a 1959 Plymouth Fury with a 361 cubic inch V-8 and was the *de facto* leader of the sailors assigned to Supply Division.

"Man, don't worry about all this heat," Squeaky said. "A Norther is coming down tonight. I bet the temperature drops thirty degrees."

Squeaky had to be kidding. I stripped down to a pair of white boxers and a T-shirt to match and waited for the night, but by eight o'clock you could hear the wind rattling the windows. The icy blasts swept down the plains straight from the Canadian tundra and when I stepped outside, I could see my breath in the air.

South Texas is a weatherman's Fantasyland, though, and by the weekend the Subtropical Zone returned long enough for us to pile into the Fury for a day at the beach. We crossed over the causeway and headed out onto Padre Island where Squeaky floored the accelerator and drove far to the south until there were only flocks of sandpipers on the beach and an occasional curlew. He slid the Fury to a stop just beyond the reach of the shore break.

While the rest of the guys tried to body surf in the confused and poorly defined waves that swarmed toward the Fury, I walked alone to the south on the hard-packed sand where the beach disappeared in the sea haze in the distance. The water in the shore break looked like the dye they use for Army fatigues and it didn't feel much like the Caribbean. I kept thinking about a book I had seen in the library back in Pacific Grove just before I enlisted, an account of a sailing trip across the Atlantic in a thirty-foot plywood trimaran. A half-mile to the north there were some drunken sailor friends and a Plymouth Fury prostrate in the sand, but I stayed off to myself and thought about Art Piver and his homemade boats. A fifteen-knot breeze swept in from the east and the sea was flat on the horizon and I thought a thirty-foot Nimble would fairly fly across the Gulf. I wondered how difficult it would be to build one.

I worked in the Training Aids Division where we maintained the ejection seat trainer and kept the film projectors operable. The days slinked by in an endless run of dawns and dusks punctuated by nightly trips to the enlisted men's club and weekend forays to South Padre Island. My own duties were light but I kept my eye on the calendar just the same. I was headed for Memphis and a series of electronics schools and a permanent duty station beyond and so there was always a sense of transition. I liked South Texas and the white sand beaches, but I didn't want to stay. An air of anticipation set me apart from the rest and I never settled into the routine established by the sailors who lived in the Supply barracks.

Just before my time at NAS Corpus Christi came to an end, Squeaky and Branson came by my cubicle to announce another road trip out to Padre Island. I had weekend duty and couldn't go. Todd went instead, as did my friend Seismo, an overweight Cajun from Baton Rouge who carried with him a cooler filled with Coca Cola and Seagram's Seven. With Jordan, Stanski and Rudy already set to go, the Fury was packed again for the trip down the sand of South Padre Island.

The boys spent the afternoon drinking and flopping about in the surf until the beer and the Coke and the Seagram's Seven were gone. On the return trip, Stanski told Squeaky to see what kind of splash they would make if they drove into the shore break. Squeaky turned the wheel to the right at sixty miles per hour and the Fury plowed into the surf throwing skyward a huge plume of sea water. There were some expensive detonations followed by the thunder of cascading foam as the Fury rolled into the breakers. Squeaky couldn't restart the engine so they bailed out and waded ashore where they turned to see the roof and the tailfins nearly buried in white water. They didn't get back to the highway to find a ride to the base until long after dark.

On Sunday afternoon, Squeaky and the boys returned to the beach in a Supply Division pick-up with a length of 3/4-inch line to try to drag the Fury out of the combers. They drove south on the sand well past where Squeaky thought his car would be and then they stopped and looked out into the foam churning toward the beach. The Fury had rolled downslope with the tide and was lost, submerged somewhere in the Gulf beyond the surf line.

When the guys wandered back to the barracks, Squeaky sagged into a chair where he sat tired and subdued. I poured him a cup of coffee from the urn we kept going on the second deck of the barracks. He took the cup with two hands and stared to the north across the bay toward the horizon and beyond where the Great Plains swept across central Texas, north toward Junction City, Kansas, where he lived before his time in the Navy. I stood nearby with my own cup and looked beyond the low buildings next door toward the horizon that appeared faintly blue. The color reminded me of the sea. The Gulf Coast looked a lot like Myrtle Beach and I stood next to Squeaky thinking about the shrimpers that made their way out of Aransas Pass. They used to drop their doors and nets close in along the coast. I sipped the scalding coffee and wondered whether any of them would ever snag a '59 Plymouth Fury.

I left Corpus Christi shortly after the Fury sank into oblivion and I never knew what became of Squeaky or Branson, or Jordan, Stanski, and Rudy, or my buddy Seismo of Seagram's Seven fame. They were the best of friends, but they were sailors first and so they disappeared into the uncertainty and the rage and turmoil of the Vietnam-era military machine.

I read a lot of books in the library on the Navy base in Memphis, including those again by Arthur Piver. There were times when I sat in the deep sofa in the back and thought about the long runs of hard-packed sand down along the barrier islands in South Texas. For a good part of the year the Gulf lays flat with those soft breezes out of the east. One of Art's little Nimble trimarans was perfect for ghosting just offshore in the heat and the stillness of a summer day, perfect for the kind of lazy sailing where you might find, somewhere beyond the surf line, the shadow of a rusted-out hulk with tailfins.

CHAPTER THIRTEEN

Sun Flower... Two words...

When I stood on the roof of the house we rented in Imperial Beach and looked in the direction of the ocean, I could see the Los Coronados Islands in the distance and to the south, the Mexican border. Just out of view, the Tijuana River emptied into a mile-long estuary that was a haven for sea birds and looked very much like the swash in Myrtle Beach where the catfish kids used cane poles and worms threaded on bream hooks to catch mullet that jumped along the channels. I didn't much care for the Southern California water temperature, though, or the winter storms that made their way south from the Gulf of Alaska, but during those summer afternoons when the northwest winds freshened, you could smell the salt in the air in our back yard. Sometimes I had to stop work and sit with a cup of coffee and let that salt air wash over me.

There is some magic involved when you can take a stack of flat plywood and build something round and three-dimensional, so smooth you want to run your hands over its surface again and again like you were a kid and you first saw the breasts of Venus in marble. As tactile and sensual as a hull might be, though, it represents only about ten percent of a completed boat. When you realize your labor of love is going to drag from months to years, the love turns into an agonizing one-foot-after-another ordeal of patience, persistence, and absolute frustration. There are countless days lost in the pursuit of lumber and screws and bronze and stainless and fiberglass, and countless more lost in abject depression because of the overwhelming number of things left to do that you can only sit in the middle of all the wood and hardware and just cradle your head in your hands and wonder in confusion how you ever got into this mess in the first place. You are gradually reduced to rejoicing in tiny victories, so tiny and emotionally draining that it might take an entire day of cutting and fitting just so that you glue and screw into place one small wooden cleat that no one will ever see. You have to keep telling yourself that if you glue and screw enough cleats, there won't be anymore left and you can move on to another endless chain of insignificant victories. After some years of all these tiny victories, you can stand on the foredeck of your unfinished boat and look across the roof of your house and see the Los Coronados Islands again and you can breathe the salt air brought in fresh daily by the cold northwest winds. For the first time in three long years you realize the end might be near.

There was a night in the middle of all those years when I couldn't sleep because the air smelled like rain and I hadn't covered the boat. In the endless cover and uncover battle with the blankets, Janie woke up beside me. Even in the dark I could tell she was smiling. She grabbed me around the neck and in her giggling, sleepy-time voice she just said, "*Sun Flower*...Two words," and our boat had a name, just like that. Another tiny victory, but from that night forward all the little victories seemed to mean a bit more. What we had in the back yard was no longer an albatross but a swan on the wing with a name so happy it even defied the cold Pacific booming a mile to the west. I couldn't repeat her name without smiling.

For months *Sun Flower* sat on her cradle in the back yard, dark and faceless, while we tried to wrap up the eight million small jobs that had to be dealt with before painting, but there came a day when Janie and I were nearly finished. Because of her name, we elected to paint the hulls white with a broad yellow stripe just above the waterline to match the yellow decks. When we stood back to look at her for the first time, she looked like a Kansas sunflower. The next day I finished installing the portlights as a surprise for Janie because she wouldn't return home until after dark. I left a droplight burning so that when she rounded the corner she could see that *Sun Flower* finally had a face. I walked alone across the street to get a better view and when I turned in the falling darkness to see the light shining through the ports, I suddenly couldn't comprehend what we had done over the preceding three years. I couldn't comprehend how beautiful *Sun Flower* had turned out to be. I sat by myself on the curb, blinking to keep away the tears.

Pictures were taken on launch day and when I look at the old albums with their fuzzy images, there is a bow shot of *Sun Flower* and all of the family and friends crowded together in front. When my great friend Greg composed the scene, I kept thinking how silly all of this was and that we just needed to get on with the launch and start partying, but when I look at the picture today, it is a remarkable record of the end of a journey, not the beginning of a dream. It's hard to look at the photograph without remembering in a flash that *Sun Flower* cost us $18,000 to build in the early seventies and took over three years to complete and that somewhere during the middle of that time, I lost two fingers in a table saw accident and that it took three months to recover enough to get back out there and use that saw again. At other times during those years Janie and I went through some severe marital tension and were even separated for a month, and still today when I look at the laughing faces in Greg's picture, I am reminded of my own failings, wading like a Zombie through that incredible building process by drinking too much and alienating friends and family because I never believed for one minute that I wouldn't finish. There was nothing else that mattered. Now I can look at the photograph and see in the faces what I needed to see in myself, that the end of the three-year odyssey marked the time when I could begin to live all those dreams that started when the catfish kid stopped to watch that schooner, a time to think again about all those shrimpers on the horizon and where *Tamarit* had gone and what it was going to be like running off before the Caribbean trades in a forty-foot Norm Cross trimaran. I was so tired at the end it seemed like I didn't have anything left.

I worked at a heavy equipment rental yard and so I borrowed a driver and a dump truck along with a big trailer owned by a couple of boat building friends. With the help of all those people in Greg's picture, we launched *Sun Flower* at the new marina in Chula Vista just as the sun went down. In those wonderful and emotionally charged moments when you know a long and tough journey has ended, you kind of float in a zone a half-step removed from reality, but there are moments that seem to stay locked forever in freeze-frame. To this day I can feel how it was to be on *Sun Flower* drifting softly away from her cradle, so light I thought of bougainvillea petals floating in the wind, and how I looked for the first time at the grotto beneath the wing decks and how the water against the hulls seemed so much like another cradle. I went below to check for leaks and I can still feel how warm and soothing it seemed in the interior. I suddenly knew, just then, I had been gone for years and finally come home to rest.

CHAPTER FOURTEEN

no cure…

A wind that blew hard all night long has played itself out. The morning is still, like your breath in a first kiss, and you can see forever the blue in the sky. One of the kids who works the counter here has propped the door open. The quiet of the morning outside slips in to mix with the quiet of the coffeehouse, and it's like listening to silence in stereo. It's that soft time after dawn when the light is right, and it makes you wish for a camera and a Harley and a long run of asphalt into the hills. There is no cure for wanderlust.

Yesterday evening I brought my little girl with me to the coffeehouse. Even though she would rather play seven innings of fast-pitch softball, she likes to sit here. While I watched her attack a *palmier* and a hot cocoa, she told me how much she wanted to go back to that place in Mexico where the little white puppies lived in the shed at the fish camp and where the only footprints left in the sand near the lagoon were ours. A young couple sat nearby talking about exotic places they had been and so I asked Marielle if the lagoon were her favorite place in all the world. She looked up and winked at me.

"No, Daddy," she said. "The Baths on Virgin Gorda," and then her face was covered with palmier once again.

The couple at the next table didn't hear. They were talking about Hong Kong and Kowloon and shops filled with jade and how mind boggling the crowds had been in the street markets at night. It made me wonder how it is that we define the word "exotic". Does a place only have to be far away to qualify, or does there have to be some kind of gaping cultural gap when you get there?

I have a good friend in Portland who drives a first-generation minivan. When an old man ran his car into it at a street corner, Bob decided the new dents along the side were only cosmetic and so he used the insurance settlement to buy a round-trip airlines ticket to Cambodia because he wanted to go see the Hindu ruins of Angkor Wat.

For most of us a destination doesn't get much more exotic than that, but I had another friend who walked the length of Baja California because he didn't want to see the landscape bouncing around from the inside of a Jeep. When he returned to San Diego, he didn't talk much about drinking beer in the cantinas of Santa Rosalia or Mulege or La Paz. He only told me about the endless nights in the desert alone when the air was hot and still and he could hear the sea birds aloft and that hollow sound from the water that lapped at the sand. The emotions he felt at being so close to the earth and to the sea made him a different man. The desert nights didn't haunt him but only slipped quietly into his consciousness when he was sitting in a car stuck in gridlock on the freeway. During those moments, he understood fully the magic and the mystery and the power of what he had done. For him the word exotic didn't mean distance or difference but, rather, a state of being. Maybe his view of wandering wasn't much different from that of my little girl who remembers, even in a busy coffeehouse, what it was like to be in a place filled with boulders and grottos and deep, scratchy sand and toasty-warm water you can see through forever.

There is a lot of quiet time at sea, but you can always hear the rush of the water beneath the hull and the occasional squeak somewhere in the rigging. You get used to the sounds and when one of the crew speaks aloud, the voice doesn't belong with the sounds of the sea and so you talk softly and laugh and then you listen again for the sea that slips by in silence.

During a night crossing in the Gulf of Nicoya, the moon went down and the sky took on the color of lampblack. A warm wind blew soft off the land and drove *Sun Flower* quietly into the darkness. There was a good place on the foredeck to sit so I leaned back and watched the masthead light carving signatures into the night sky while I listened to the rush of water off the bow.

I was beginning to nod off when I heard a dull pop and an exhale and a squeak. I crawled forward to lie on my stomach beneath the rail and to look into the water that was as black and depthless as the night. Two dolphins appeared suddenly. There was so much bioluminescence in the water they were backlit in green. More of the green cascaded from their beaks to flow back toward their flukes and then it trailed away, like green fire trailing from a pair of F-18 Hornets at an air show.

The dolphins played in the waves off the bow while they squeaked and banked to the left and then squealed and banked to the right and then they peeled away and disappeared into the depths trailing green fire into infinity. Out of nowhere, the green trails appeared again and got larger until the green burst through the surface where the two of them popped and exhaled and squeaked again. The green exploded all around and I could only lie quietly beneath the bow rail and watch. When the dolphins tired of the game, they squeaked again and broke away and in the end, the trails of green fire disappeared into the depths. There was nothing left again but the rush of the sea and the masthead light scribbling away at the slate overhead.

I don't often think of that night with the dolphins, but at odd moments when I'm hot and tired and irritable and I'm sitting in gridlock on the freeway like my friend who walked the length of Baja, the dolphins come back to me trailing their green fire. For a moment there is the rush of the sea and the dark of the night and I wonder to myself how I ever let them go. There is no cure for wanderlust. Not even wandering.

CHAPTER FIFTEEN

blue neon...

The neon sign in the back of the Kensington Coffee Company has very much the same color as the shallow water in the Caribbean. While I waited for my French roast to cool a little, I kept looking at the sign and it made me smile. I spent a lot of time wondering about the color when I was young because at night you could see the blue neon of the marquee over the Gloria Theater reflected in the windows of the cars that passed in the street. I leaned against the popcorn machine and watched the reflections flashing by, but sometimes I walked out onto the sidewalk just to stare down the hill toward the sea that looked dark as the night sky. There were pictures of the Caribbean in the library and I looked out over the dark Atlantic thinking about the color of the water in the pages of *National Geographic*. I always wondered if the magazine retouched the photos because the colors were the same as the blue in the neon signs over the theater that kept flashing in the car windows.

I wanted to see the colors of the Caribbean firsthand. After we sailed into the muddy harbor in Punta Arenas, I lost my patience. I didn't want to wait until after we transited the Panama Canal to see if the pictures in *National Geographic* were untouched. We left *Sun Flower* in the hands of friends and took a short train trip over to Puerto Limón on the eastern side of Costa Rica.

The bus from Punta Arenas took us to San Jose where we discovered a lovely old city with an opera house and narrow streets and statues in the plazas and flowers in the park and a busy station where we boarded a narrow-gauge railway for the trip over the mountains to the east.

Once the train left San Jose, it began the long ascent through the interior of the country and over the hills toward Limón, chugging like the tugs that pushed the barges down the Waterway when I was a kid. In each of the villages, we came to a stop with a whistle of steam and a squeal of brakes while the train was boarded by dozens of street urchins selling papaya and coconut and mangoes and local tamales of rice and corn and pork wrapped and steamed in banana leaves. Each time the train whistled and squealed to a stop, we bought more tamales because they were good and fresh and because Janie thought all the street urchins were darling. When the train began to move away from the towns, some of the kids jumped early on while the bold ones held their grips until injury or death was imminent, but all of them hit the ground running and laughing and waving at the passengers who were busy eating the tamales.

In the evening the train descended from the hills into the low area west of Limón, chugging its way toward the Caribbean along tracks lined with tiny, neat houses sprinkled with orange flower petals falling from the *flamboyans* trees. The sinking tropical light saturated the colors and made the Lowcountry turn lush and thick and green.

Limón was quiet and quaint but when we stepped from the train, Janie and I were greeted with the quickening drops of an evening downpour. We hustled our way down the street where we rented a two-dollar room in a *pensión* near the square. When the rain stopped just at twilight, we walked from the pensión to catch our first glimpse of the Caribbean, but the tropical darkness fell like a drop cloth. When we got close, there was no neon blue left to see.

Neither of us felt hungry after the tamales on the train. The market was still going so we only bought cheese and bread and walked back to the square where we sat on a bench and ate in the heavy night air. Our roommate was snoring when we crept into the pensión, but the sheets were clean and the fan kept us cool and there were no mosquitoes to slap. The night passed quickly we were so tired.

Light from the single window woke me at sunrise so I washed and dressed near the sink down the hall. Janie followed soon after and we left the pensión while our roommate was still snoring. We crossed the square where a man in a kiosk was selling coffee that he poured for us in eight-ounce water glasses. The coffee was so thick and strong your cheeks shuddered even after you cut it with an equal amount of cream, but the flavor was the same as the aroma you get when the coffee beans flow through the grinder like dust. Janie and I had another coffee and sat once more on the bench in the square.

The morning was overcast with sections of the sky nearly black. The rain was coming again. We knew the Caribbean would be gray to match the sky and so there would be no neon blue to see. The market opened early, though, and we strolled through the stalls and bought more bread and cheese and looked at all the fruit. We found another coffee man who told us we should ride the bus back to San Jose because it followed a narrow dirt road up and over the top of Costa Rica and it was a ride not to be missed. We looked for the little bus station where we bought our tickets for the ride not to be missed and we left Limón behind at mid-morning, bouncing around in a miniature Mercedes bus with the skies still overcast and the Caribbean lurking gun-metal gray beyond a hundred tin roofs.

The coffee man in the Limón market knew a good thing. As the Mercedes climbed the dirt road in first gear, Janie and I stared through the windows at the rain forest sliding by and at the peaks looming in the distance. The road was dry, but the bus slipped and slid its way through potholes and up the switchbacks. We passed houses with yards full of chickens fluttering about and dogs panting in the dust the bus left behind. When the rains came it was only a light sprinkle that helped keep the dust down and we climbed slowly toward the clouds where we stopped for a stretch in a village near the top of the pass. There were more darling street urchins selling tamales so Janie and I bought two apiece and ate them standing beside the road, looking back toward Limón where the view from the mountain was only partially blocked by the rain toward the coast. After the happy passengers had all boarded, the driver beeped his horn and we bounced our way toward San Jose watching the chickens again through the windows of the bus.

We reached the narrow pavement that led into the city where the bus driver accelerated through the traffic and kept Janie and me at the edge of our seats. The locals never noticed the speed and when we reached the station, all of them left the Mercedes with smiles and laughter. We were just happy to get off the little bus and into a coach the size of a Greyhound. The new driver wasn't much slower. He drove through the hills like Juan Manuel Fangio, but we got back to Punta Arenas in time to stop to buy some rum for evening cocktails.

Sun Flower floated lightly in the muddy water of the inner harbor. After we hailed our friends to bring the sabot over, we sat at the edge of the landing and waited. I kept looking at the dark water and at the mud that lined the channel. I was sorry we had missed the neon blue of the Caribbean. Janie handed me the rum and I took a swig and let the alcohol burn its way to my stomach. I liked the train and Limón and I liked the street urchins and the tamales and I liked the coffee in those water glasses in the square by the pensión. Janie took the rum back and I winked at her while she took another sip. I kept thinking about the Caribbean, though, still wondering if it really was the color of the blue neon tubes on the old Gloria Theater marquee.

CHAPTER SIXTEEN

into the shadows...

When the door of the Starbucks is left open for a moment, you can hear the cars on the freeway beyond the shopping center. The sound they make is a soft rush, like a morning wind, and it reminds me of sitting in the sand in Myrtle Beach where the sea oats grew tall. I used to sit in the dunes and listen to the wind and to the sound of the sea as the waves boomed and then rushed and then hissed along the beach.

An old man came in a few minutes ago. After he settled in with his coffee, I saw he was missing part of his jaw. I watched him for a moment while he cupped his hands and sipped from his mug and when he swallowed, he tipped his head back slightly so the coffee wouldn't spill from his lips. I was embarrassed, suddenly, when I realized I was staring. I looked away, but not before he caught my eye. It may be that he was used to the stares of strangers because there was a slight off-center grin on his face. I nodded and smiled and turned my face to stare through the window instead.

In Panama I was told to stay away from the Guardia Civil, but the Guardia and their AK-47's were on every street corner. They stood alone and in pairs dressed in uniforms that were almost Nazi, with polished boots and polished holsters, but their caps were silly affairs that swept upward like Venetian gondolas. A member of the Guardia had a post near the yacht club in Colon where I kept *Sun Flower*. He watched with interest as I went about provisioning the boat for a long, wet slog to Jamaica. He didn't try to detain me nor did he speak, but he stood rigidly when I was near and he kept his body positioned so that the AK-47 he carried was in full view. I didn't trust him.

A well-dressed Panamanian kid of about twenty hung around the yacht club as well. When I had business in the *aduana*, I saw him glancing my way just beyond the office doors. When I went into town to have work done on the starter for the diesel aboard *Sun Flower*, I saw him again on the street outside the shop. He reminded me of a shark circling at the edge of the gloom, following me from a distance at any number of places when I was away from the docks. I thought he was an informant of some kind and I didn't trust him any more than I trusted the Guardia Civil.

Janie had lived in Panama as a child when her father was stationed on the Navy base across the bay from Colon. She was too young then to remember much and so her memories of Panama were just sketches, but she had aboard *Sun Flower* a couple of wrinkled, black-and-white photographs of a Panamanian local everyone knew back then as T-bone. He had been a family friend and a constant drinking companion for Janie's father who drank like a sailor on permanent shore leave, but when the family was transferred, they returned to the States while T-bone stayed in Panama and was swallowed into the shadows of distance and time and poverty. Janie wanted to find out what had happened to him. On our visits into town we asked the storekeepers and taxi drivers about the man named T-bone. The people only shrugged their shoulders.

A few days before we left to sail to Jamaica, the kid who acted like a police informant began hanging around the yacht club again. I should have questioned him about his spying, but it seemed harmless enough to ask if he had ever heard of a man named T-bone. There was some hesitation and then he looked at me.

"Why is it you wish to see this man?"

"It's for my wife. She and her family knew T-bone when she was little."

Janie brought the old pictures out to show the kid. He stared at them for a long time.

"Yes," the kid said. "I know him."

In the afternoon I heard a knock on the hull of *Sun Flower*. When I slid the companionway door open, an ancient black man stood alone on the dock. He was thin and angular and stooped, and when he turned to look at me there was only sagging flesh where the right side of his jaw used to be. Janie stepped out into the cockpit and you could see the tears glisten in her eyes. She hopped over the rail onto the dock where she giggled like a second-grader. The two of them stood near the lifelines and held each other.

The three of us sat together below on *Sun Flower*. There were a lot of stories to share and T-bone would tell them slowly at first because you could see he was self-conscious about his appearance, but then Janie would laugh and T-bone couldn't help himself and he would try to cover his face with his hands. He laughed so hard he couldn't hide anything and it made you forget all about his jaw.

I had an uncle in Arkansas who ate limburger cheese and who laughed a lot, but he always chewed tobacco. When someone said something funny, he thrust his lower lip forward so the juice wouldn't spill and it gave him a laugh just like T-bone. You had to love the old man for it. T-bone thrust his mangled jaw forward and rocked back and laughed and he sounded just like Uncle Gus. I laughed with him and with my wife and there was a feeling that day of a kind of timelessness, that the years they spent together were priceless and not replaceable and because they remembered those years and laughed about them, they were now real and permanent and alive and everything was going to be okay because all of us knew the old man wouldn't live much longer.

Late in the afternoon when T-bone had to leave, I walked with him to the end of the dock where there was a lot of soft sand. The informant kid came to help.

"Thanks," I said to him. "Thanks for finding T-bone."

"Your wife. She's happy?"

"Part of her childhood came back."

I shook the old man's hand and then we hugged for a moment. I stood in the wind that blew steadily across the water from the east and watched as T-bone limped his way toward the Guardia who stood near the yacht club entrance. I saw the soldier stiffen in the wind and I got the odd feeling he was standing at attention. When the old man passed close to him, there came a moment when the soldier bowed slightly from the waist. T-bone walked past him and held his head high and then he disappeared into the street beyond. I stood near the docks and looked toward the soldier. There was no sound except for the wind and the slapping of a loose halyard on a boat behind me. The Guardia turned to face me and then he straightened himself again and nodded in my direction.

The man in the coffeehouse who had no jaw has disappeared into the parking lot. I'm still here with the cold coffee in the mug, listening to the cars on the freeway that sound so much like the sea in Myrtle Beach, like the trade winds that blew across the bay in Colon. I have a vision of T-bone again walking across the yacht club grounds with the wind whipping his shirt and the outline of his wishbone legs pressed into the back of his pants. It would be great fun to laugh with him again because you had to love that old man, but when he walked away across the sand that day, he stepped back into the shadows of time. Janie and I never saw him again.

CHAPTER SEVENTEEN

Jamaica then, my friend...

Down in Jamaica's Port Royal, a small hotel with a dock overlooked the channel that led into the big harbor of Kingston. The hotel management didn't mind if cruising yachts tied up for a day or two while the crews refilled water tanks and took on supplies. The dock had an outside bar and a freshwater swimming pool nearby and it was a good place to meet people who wanted to sell you Jamaican currency for American dollars without a lot of concern for the official exchange rate. In the mornings it was easy to sit and drink cups of Blue Mountain coffee while you sold your cash and watched the day unfold.

I spent a lot of Jamaican currency on the coffee and on the good Appleton rum. During one of these visits, I sat alone with a rum and tonic and watched as an older wooden yacht tied up alongside. The owner and his guests were laughing and singing and plopping onto the dock where they walked over to order drinks from the bartender. The owner of the boat turned to me and offered a refill of my rum and when I accepted, there was a lot of laughter and backslapping and suddenly I was caught up in a party that didn't end until I sailed away from Jamaica five days later.

The owner of the yacht was a medical doctor who worked at a hospital in Kingston, but he was from an older and more conservative generation that was confused by the marijuana culture and by the social unrest that swept through the back streets and the shantytowns of mid-seventies Jamaica. We talked about these things and you could see the good doctor was struggling with the politics that forced nighttime curfews on the people of the streets and kept the tourists away. He was convinced Jamaica would recover and the old ways would return.

"Ah," he said. "You should have seen Jamaica then, my friend," and he would look to the mountains above Kingston where the clouds tore away to the west. You were swept away as well because, for a moment, you could see Jamaica through the doctor's eyes.

The next day a messenger from the hotel came to *Sun Flower* with a request that Janie and I join the good doctor for a dinner party at his house. There was a note attached saying we should arrive several hours early because there were some things the doctor wanted to share. We hired a driver who took us through downtown Kingston and then into the hills above where the houses were big and the roadways were quiet and there was none of the tension and meanness we felt on the streets lower in the city.

The doctor met us at the door where he swept us in with a great flourish and then he poured shots of rum with tonic and lime and then took us to his study at the back of the house. Taped to the wall over his desk was a profile drawing of a big racing yacht called *Xayamaca*. I looked at the lines and thought immediately of the great *Windward Passage* because of the long flat run aft. *Xayamaca* could be just as fast off the wind. The good doctor stood proudly in his den.

"She looks like a bloody great dinghy, don't you think?" and he stared with pleasure at his drawing on the wall.

With a lot of enthusiasm he explained that Jamaica needed something like this, a world class grand prix yacht to campaign in the long ocean races, a yacht built of native hardwoods that would show the world what Jamaicans could do, but then he stopped himself short and he looked at me and shrugged his shoulders.

"Jamaica is falling apart, I'm afraid," he said, and he looked back for a moment at the drawing of *Xayamaca* and then he turned out the lights.

We followed the doctor out to his driveway where we climbed into his Rover. He drove us slowly up into the hills beyond the city where we parked near the entrance to the botanical gardens. We left the car behind to walk among the walls and trellises and statues streaked with bird droppings and to climb the hill through the weeds and the vines that had invaded the sanctuary. There were some red and yellow hibiscus plants in bloom and the shrubs that had gone wild were green and thick and healthy, but the orchids hidden in the shadows were stunted and woody and overrun. There was nothing much left of the gardens.

When Janie and I scrambled to the top of the hill, we turned to look beyond the walls and the empty fountains to the Caribbean below, sweeping in blue neon toward Panama far to the south. There was something about the sea then, something clean and pure, that made the botanical gardens seem like so many people we had seen in the streets: old men who were dirty and old women who were neglected, and all of these old ones had the blank stares of people who were going nowhere. When the good doctor joined us at the top of the gardens he, too, looked away to the sea and then he spoke in a quiet voice.

"It is the view I wanted you to see," he said, "but the gardens were beautiful once, as you might tell."

He looked again at the weeds and the tangles of vines and at the bird droppings that streaked the statues and at the fountains that were silent in the early evening air.

"Ah," he said, forcing a smile. "You should have seen Jamaica then."

CHAPTER EIGHTEEN

the Zen of his own coffee...

I first saw the ship as it emerged from over the horizon, watching it intently as it closed with *Sun Flower*. I wasn't sure we were on a collision course and I kept tracking its movement as it approached, but then the rain started. I lost sight of the ship and there was nothing to do but wait. I sat in the cockpit and tried to make myself invisible by hiding in my foul-weather gear. A leak around the collar allowed the water to seep into my shirt and it made me want to shiver even though I was hot in the jacket. The sound of the rain thundered on the deck. I began to worry about the ship and where it might be and whether I would hear the throb of its engines, or would it just explode from the gloom and run me down and crush *Sun Flower* deep into the abyss. I couldn't do anything and so I sat there steaming and shivering in the foul weather gear, trying to scratch the tickle from the leak and listening for the guttural beat of the engines. I thought about the music from the movie "Jaws" and how discordant it seemed and that the systolic beat was all tension and no resolution. I sat in the cockpit nearly in a Zen trance, barely breathing, straining only to hear through the rain the discordant sound of the ship.

As the minutes trickled by, the rain increased its tempo. It came down so hard it beat the sea flat and stole the life from the wind. Through the thunder on the deck I could hear the sails slapping and crackling and there was a squeak in the gooseneck when the boom flopped about. The boat shuddered as the mainsheet traveler slammed against the stops. I sat motionless in the cockpit and stared into the gray sheets of rain.

I remember thinking about the recurring nightmares I had as a child where there were two railroad tracks that crossed at right angles. A locomotive was on each of them. Always when the nightmare began, one locomotive crossed slowly at the intersection and then in a few minutes the other came chugging through. As the nightmare continued, the trains reversed themselves and picked up speed to cross the intersection, over and over again, and they got closer and closer to a collision. I dreamed this was happening in the dark when I couldn't see and I rolled and kicked in my bed as the locomotives shrieked through the night, faster and faster until at the moment of impact I awakened in the dark, soaked in sweat and screaming into my pillow.

The memory of those nightmares made me shrink like a kid again into the foul weather gear. I stretched the hood to cover more of my face, but then I began to hear the freighter. A throb from somewhere sounded like the nightmares again, only I couldn't place the direction of the sound because it felt more like a deep heartbeat. I wasn't sure of the throb except the beat was getting heavier, like feeling the pulse of a ship instead of hearing it. I strained to see through the downpour. There was nothing out there except the thunder of the rain and the gray and the growing throbs of the engines.

A few weeks before, Janie and I had gone diving on a reef down in Panama. I wanted to check the set of the anchor first and when I swam from the boat to catch up to her, I watched through the clear water as she moved with lazy kicks toward a coral head off in the gloom. I was about thirty feet behind when I saw from the corner of my eye a shark swimming fast through the water and headed straight for my wife. In that moment of terror when you suddenly know you are helpless, I saw the shark would cut her in half it was so big. I screamed for her to turn and I choked and gagged in the salt water and when the shark was nearly upon her, I could feel my head throbbing with the screams I wanted her to hear. There were no sounds in the water beyond my own gagging and choking. Janie kept up her lazy kicking toward the coral head while the shark zeroed in on her midsection like some kind of primordial Sidewinder intercepting its target. I was cringing for the impact when the shark suddenly veered away and swam out of the picture into the gloom. Janie never saw the shark. She didn't know when I caught up to her why I was shaking.

There was nothing for me to do again but sit in the cockpit and sweat and shiver in helplessness. I thought the ship would explode out of the wall of rain with a blinding crash, like the trains that collided in my nightmares, sending me spinning and retching and gagging into the cold silence of the sea. I sat in the cockpit and waited like the little boy who couldn't stop the nightmares, unable to stop the throbbing of the engines in the freighter. Rain pounded the deck while the pulsing of the ship surged into my soul like visceral drumbeats. A huge mass of steel, shadowy and cold and murderous, morphed from the rain to pass in front of me. I watched as it thundered by and thundered by and thundered by, and with the last of the thundering the ship disappeared once more, swallowed into the dark, shivering wall of rain.

The pulsing of the engines began to fade and I became aware of the pounding on the deck again. The throb of the nightmares was going away. I shivered again and thought about that big shark and how it would have cut Janie in half. I wondered if the master of the freighter had been standing on the bridge drinking coffee and peering into the rain as he went by. I wondered if *Sun Flower* was a blip on the radar screen, or was there nothing? Maybe, while I was being swept into the nightmares once more, I would have been there one minute and then have been gone the next. Maybe the captain sipped his coffee and glanced at the blips of the other freighters in the sea lane and then took a deep breath and wished for the rain to stop, but it wouldn't have mattered to him because there was nothing for his ship to hit for miles. Sitting there alone in the cockpit of my boat, I would have disappeared into the thundering rain and into the flattened sea. The captain of the freighter might have known nothing but the Zen of his own coffee.

When it was over, I breathed a huge sigh of relief. I felt my lungs quiver like they do when you are a little boy and you've been crying. The deep pulse of the ship had receded into the sound of the rain and I heard again the squeak of the gooseneck and the banging of the mainsheet traveler. I thought once more about the captain on the bridge of the freighter sipping his coffee, staring at the green blips on his radar screen. I reached around and rubbed the wet part of my neck and stood and stretched, but I couldn't stifle a nervous yawn. Beyond the gray of the rain that beat on the deck of *Sun Flower*, the freighter throbbed its way toward the Yucatan Channel. I looked into the gray to the west and shuddered.

There were cat's-paws at first. The rain fell hard but not so thick and then the wind began to fill from the east. I steered *Sun Flower* onto a close reach that carried her toward the Dry Tortugas. After she eased up to five knots, the rain fell away to nothing. I glanced back to the west after the rain cleared away and the hull of the freighter was nearly lost over the horizon. All I could see was her white superstructure and then a squall moved through in the distance and the freighter was gone. I wondered again if the captain were still lost in the Zen of his own coffee.

CHAPTER NINETEEN

perfect circles...

I stopped for coffee this afternoon at a place where there were tables outside and a handful of yellow umbrellas. The café was clean and bright inside so I bought a cup of Italian roast and then sat down near the front. While I waited for the coffee to cool, I stared absently through the window at a girl outside who was sitting with her scruffy boyfriend. They were talking and laughing with someone who was out of my view, but I kept watching the girl. She was a living Calvin Klein ad except that she had very large breasts. I looked at her for a moment and then stirred my coffee. When I glanced up again, she was walking away hand in hand with her boyfriend. I kept thinking about her and how she sat smiling beneath the yellow umbrella.

Pirate Jud was a great friend of mine in Key West but when people saw him around the docks, they thought he was armed and dangerous. He had the size and build of your average heavyweight and his skin had taken on the color of a worn razor strop from all of the years in the sun. He didn't wear shirts, even when he went to town, only a leather vest polished smooth from constant use, and always he had on a pair of faded cut-off Levi's. He started shaving his head years before and a lady friend in town who owned a boutique decided he needed a knife to wear. She pierced his earlobe and sold him a tiny silver dagger that followed the curve of his ear. It used to glint in the sun and remind you that he was not someone to be trifled with. He was a good family man, though, who lived with his wife and his three daughters on a modest ferro-cement cutter in a marina out on Stock Island. There was money to be earned doing odd jobs around the docks and Army disability checks to collect each month and his wife drew unemployment and qualified for food stamps. The five of them shared with everyone what little they had.

I lived on my boat only a few slips down from Pirate Jud. Once in a while late at night he would come down the dock and knock on the hull of *Sun Flower* to tell me he had to have company on a trip into town. I didn't get out much except to drink beer at the Green Parrot Bar and Sub-Shop down on Whitehead Street, but Pirate Jud only enjoyed the topless bars. There weren't any out on Stock Island. We took the bus into Key West where there was a club just off A1A that would pay a good hourly wage to any girl who would dance and expose her breasts. We never knew who might be on stage, but after a little drinking it didn't matter much. One night I glanced up from my beer and discovered the girlfriend of a guy I worked with at Key West Hand Print Fabrics had removed her top and was dancing on the stage. When she was finished, she covered herself with a towel and came giggling to our table where she sat for a while, embarrassed and shy. Pirate Jud was charmed and mildly in love and didn't want her to leave to dance again.

Later in the evening there was a girl on the stage who was so afraid and nervous she couldn't move to the music. She stood there alone staring out at all the drunken men. Some of them began to laugh and make comments about her breasts, which were large and nearly conical in shape, and there was no sag to them. The impression you got was of two bulls-eyes in 3-D. Some guy kept saying how perfect her tits were and there was some more laughter and someone else said they reminded him of the advertisements for Perfect

Circle piston rings. The name stuck and while the girl stood there and tried to dance, all the men called her Miss Perfect Circle. They kept laughing at the joke and yelling at each other to check out the tits. Pirate Jud only sat in his chair. After a few minutes of the howling and the laughing, you could see the disgust in his face and he just looked at me and motioned toward the door. We got up from our table and I glanced at the girl on the stage. She stood there staring at something in the darkness beyond the men in the bar while Pirate Jud led the way to the entrance. Behind us we could hear the drunken businessmen howling at their stupid jokes and laughing at the poor little girl who couldn't dance. We walked out of the bar into the heat and into the shadows of a big August moon.

A friend of mine told me years later that Pirate Jud had tired of the tourists in Key West and had loaded his cutter with everything he owned and then headed out across the Gulf to Apalachicola. There was a hurricane approaching and Pirate Jud's boat wasn't rigged for sail yet, but he left anyway. I never knew what became of him.

The tables beneath the yellow umbrellas outside are empty. I keep looking through the windows even though the Calvin Klein girl is gone. She reminded me of Miss Perfect Circle Piston Rings, and she reminded me of one of the world's more unique human beings. I hope Pirate Jud and his family made it across the Gulf.

When my mother came down to Key West for a visit and met Pirate Jud for the first time, she was nervous and afraid. After a few days she came to understand what the locals already knew and what Miss Perfect Circle couldn't see. The only gold Pirate Jud ever had was in his heart.

CHAPTER TWENTY

trip into the night...

Voices from somewhere on the dock woke me from a deep, sweaty sleep. I opened my eyes and didn't know at first where Janie was. The thought of her lying naked next to me started the hot flashes again and I had to get up and splash water on my face. I pulled on some shorts and looked through the open companionway where I could see the vertical outriggers of a shrimper moving by in the Singleton marina next door. The sound of its diesel echoed from the walls in the basin. I stopped to listen before filling the coffee strainer with Bustelo. The kerosene pressure tank had to be pumped again, but the stove lit right off and in a few minutes the aroma of Cuban roast spilled through the cabin. I poured the first cup and added milk and sugar and then stepped out of the companionway to sit near the helm. A pair of seagulls kept squabbling near the masthead on the boat next door. The noise and the flapping around made my head hurt. I drank the coffee down and grabbed a towel and the shaving kit from below and walked up to the showers at the head of the dock.

The restroom was empty and quiet and dark. I took a lukewarm shower in the shadows and then stared in the mirror at the bags underneath my eyes while I shaved. I slipped on a clean T-shirt and a good pair of cut-offs and stepped outside and walked away, blinking in the noonday sun. Pirate Jud met me in mid-stride.

"Let me show you something," he said.

Jud led me over to the slip where his thirty-five-foot

cutter spent its time oxidizing in the sun. There were five girls in bikinis lying about on the cabin top and on the foredeck. I didn't want to embarrass anyone and I tried not to stare, but the girls were beautiful. When Jud introduced them, I had to keep moving toward *Sun Flower*. Pirate Jud walked down the dock chuckling while I slid the companionway door open.

"Don't get yourself all lathered up," he said. "Those girls are topless dancers from the bar downtown. I think they're lesbians. What a waste."

"You know, it's a good thing they are," I said. "Dottie would feed you to Wolf out there, piece by piece."

"You're not kidding. Going into town tonight?"

"Probably. I might just wander down to the Green Parrot again."

"No topless bars, huh? That's okay. All the girls are here, anyway."

Pirate Jud grinned and made his way down the finger toward the dock. I took advantage of the fine weather and opened the forward hatch to let the air circulate through the boat. Everything was in place and you couldn't tell I had been gone for so long hauling gill nets on the mackerel boat. I put fresh sheets on the wingdeck bunks and clean pillowcases and I didn't plan to sleep, but I could feel the emotions again welling up from somewhere deep. I put my head down for a moment and stared through the portlight at the empty sky. I felt like burying my face in the pillow but I could hear Margie and the kids from the houseboats running down the dock chasing a kitten. When the kitten slipped off the finger into the water, one of the shrimpers came over from his job mending nets in the field across the way. I could hear them talking.

"That kitten is barracuda bait," the man said.

"No, it's not," Margie cried. "Can't you get her out?"

"I don't know. The water is pretty deep."

Margie kept whining away until the shrimper fished the kitten out of the basin and I couldn't hear her anymore. I got out of the bunk and ran some more water over my face and went out into the cockpit to sit in the afternoon sun. All of the kids were huddled around the kitten over near Smokey's houseboat when Beth the Writer walked down the dock. She stopped to use her bathrobe to wipe the poor thing dry and I could see a good part of her legs in the light. Beth caught me staring and she walked out on the finger to say hello. I could tell she had already started drinking and it wouldn't end until she passed out. The sight of her legs went straight to the pit of my stomach. Beth stood on the finger like she knew. For the first time since Janie left, I was suddenly caught up in desire and I could feel my face heating up. The moment drifted by and I didn't invite her aboard. She put her hand on the lifelines and smiled.

"I haven't seen you around lately," she said.

"I've been back and forth, fishing for mackerel down near the Marquesas Keys. The fish are almost gone."

"Dottie told me about Janie. I'm sorry."

"It was her choice, I guess. She won't be back."

"Why don't you stop in to see me? I've always got rum and beer and things."

"I know. Maybe I will."

Beth reached over and touched my shoulder and I could feel the heat in my face again. She turned and walked up the dock toward the showers. I sat for a minute thinking about what had just happened and then Margie came over with the shivering kitten. I cradled it close to my chest until it started purring.

Sun Flower needed a good cleaning on the deck. I used the hose to wash away the bird droppings and the dust from the field next door and I cleaned the portlights. When I finished near sunset, I felt better about going into town. I even thought for a minute about Pirate Jud's topless dancers.

I washed up in the galley sink and changed into a Chico State T-shirt and then poured a sixteen-ounce glass of Carlo Rossi burgundy. The wine was warm and a little bitter, but I drank it down and then stepped off the boat and walked across the field beneath some towering thunderheads to catch the Green Line into town.

I found the corner seat in the Parrot and ordered a draft beer while some heavy rolls of thunder rumbled into the bar from not far away. In a few minutes a good Southern downpour started. Lightning and crackling shook the old building. I thought the water rushing by in the gutters might have moved some parked cars if the curbs weren't painted red. Most of the regulars paused to watch and listen from the light in the open bar. The rain came down hard and closed the place in and it felt like time itself was suspended and there was nothing left for me in Key West. I sat alone at the bar, angry and frustrated because I had let Janie go.

Beth came running in streaked with rainwater. She took a seat at the other corner where she started talking to Old Man Ned who was working that end of the bar. I thought about Janie again but I still had visions of Beth's legs shading into Neverland, and there were all those bikini girls slathered in suntan lotion and flirting with Pirate Jud. I sat there thinking about Beth and the topless dancers, wondering what I was supposed to do. Janie still called me on occasion but our conversations on the phone were laced with bitterness no matter how upbeat she sounded. I felt guilty when I glanced at Beth across the way.

I left Kellog a five-dollar bill to buy Beth a St. Pauli Girl or two and then I walked in the last of the rain up Southard Street toward Duval, still mad and upset and not sure of anything. When I passed by the Conch Tour ticket windows near Mallory Square, Deanie and Biker Bill rode up on their Hog and waved me to a stop.

"We're out rounding up people for a party," Deanie

shouted. "Why don't you come by?"

"Where's the party?"

"Over on Billy Goat Lane. You can't miss it. Everyone's coming so you might as well join us. Might be some ladies there, too, if you're up for that."

I didn't see Biker Bill and Deanie very often. She used to do Janie's hair, but that was over. I thought a biker party would be different so I walked back up Duval and turned left on Fleming to find the house.

Bailey was sitting with Franco on the front porch along with Willy Dog who wagged his corkscrew tail looking for a handout. I had an urge to beat Franco's face to a pulp. I turned to leave, but Deanie came sweeping down the steps like Loretta Young in black leather. She glared at Franco and then she took my hand and led me into the conch house and to the kitchen where there was a cooler full of St. Pauli Girl and another full of Beck's on the floor. There were shots of rum and tequila on the counter so Deanie left me alone in the kitchen to fend for myself. The tequila was good with the limes and I drank several of the shots hoping the alcohol would get me out of the funk. I drank a jigger of rum and then took one of the bottles of beer and went into the living room where there were a few locals I knew. More people were trickling in so I moved to a corner where I didn't have to talk to anyone.

Someone lit a joint and began passing it around. Deanie sat on the floor and took a big hit before giving it to Cindi who was the cashier I knew at the Pier House. Bailey came in when he smelled the dope and got in line.

"You don't have to worry about Franco," he said. "He and Willy Dog went down to Duval Street."

"I didn't think he would come in. Thanks for looking out for me."

"You both need to make your peace. He knows he made a mistake."

"Every time I see that guy I have to force myself to keep from beating his face in."

"You'll get over that. Just let it be."

Deanie turned up the volume on the stereo to hear Dave Brubeck and it was hard to talk. She brought me another St. Pauli Girl and I drank the first one down and started the second. I was already full from the Green Parrot and not feeling much from the wine and the tequila and the rum. The smoke from the joint filled the room and made me queasy.

Bailey was having a great time with it as were all the people sitting on the floor. One of the bikers took a vial of amyl nitrate out of his pocket and passed it around. Cindi removed the lid and tried to sniff the fumes, but the chemical jolt was so powerful it knocked her head backwards. No one but Biker Bill could breathe the fumes for more than a second. He handed the vial to me and I looked at it for a moment. The cap came loose in my hands and when I raised the vial to my nose, someone picked up a medicine ball and smashed me full in the brains. The flames from the lighter fluid shot up my nose and seared my lungs and stole away my breath and I could hear Biker Bill laughing down a long, narrow tunnel that kept moving up and away and I couldn't keep my face in the light to see. Bill took the vial away laughing hugely. He took a long breath through his nose in the fumes and staggered a little and laughed hugely again and then he offered the vial to Deanie who sat stoned on the hardwood floor.

I went back to the corner near the window. The moving tunnel stayed with me. The St. Pauli Girl had picked up an isopropyl taste and I couldn't drink it. I looked around the living room of the conch house and saw my friends who had been laughing so hard a few minutes earlier. It all seemed silly and pointless and I didn't belong there anymore. Biker Bill was still laughing and Dave Brubeck made the floor shake, but most of the people at the party had done too much. They were sitting around in random groups, staring through hollow eyes like little schools of tuna.

I stepped through the front door and tried not to trip into the night. I could still smell the vaporized diesel fuel that had been sprayed earlier by the mosquito plane. Willy Dog came panting back from out of the darkness and hit me up again for a handout. I scratched his ear and sent him in the direction of Franco's place. I trudged alone down the street moving vaguely toward Stock Island. There were palm fronds hanging over the fences and I could feel them brushing my face. I kept my head down and looked at the jigsaw puzzles in the sidewalk. Someone I knew honked at me and I turned in time to see red taillights receding down the street. A soft east wind kept rustling the leaves of the magnolias. There were lights burning in some windows but everything looked vacant and I was confused.

I crossed the bridge over Cow Key Channel where I could see to the south toward the Gulf Stream a bank of heavy clouds that loomed over the sea beyond the reef. There were flashes of distant lightning. The sweet smell of rain mixed with the thick salt air that drifted in from the flats, but beyond the line of mangroves crowding the channel the smell of dirt returned when I walked alone past the abandoned greyhound track. After I turned the corner at the end of the block, I could see the security lights hanging over Tropical Marine Center and the cluster of masts in the shadows on the other side of the dock. I crossed the field of dust and torn shrimp nets and looked toward the glow of the marina hoping somehow that Janie would be there, or Beth the Writer, or maybe even Josie who worked the sub shop at the Green Parrot, but *Sun Flower* danced alone in her slip. I flopped into the bunk with my head spiraling down into that same dark tunnel from Biker Bill's party. It was still full of fumes from the vial of amyl nitrate.

CHAPTER TWENTY-ONE

Sandy...

When I drove over here a few minutes ago to get coffee, I was thinking of the color of the sky and of the setting sun and having to drive into it for long periods of time and how sleepy you can get. I thought about my friend Jake who was killed down in the Florida Keys by a drunk driver who crossed over the centerline and hit his car head-on. It was just four-thirty in the afternoon, about this time of day, and Jake was on his way back to Key West from a construction job up on Big Pine. With the low sun in his eyes, he never had a chance.

Jake was German and spoke English with just enough accent to attract the ladies. He wasn't a tall kid, but he had broad, muscular shoulders and dimples that creased his cheeks when he smiled. The ladies followed him around like a line of ducklings. I nearly always sat in the same corner seat down at the Green Parrot and Jake would wander through after work towing a lady or two. We drank beer together and talked a lot about sailing. He was one of the good guys.

One afternoon Jake walked in with a young lady at his side who was a stranger to the bar. They were holding hands and kissing and hugging and when he saw me at my corner seat, he came over to introduce me to Sandy who was the love of his life. In a bit of Germanic bluster, he bellowed at my friend behind the bar to bring me another St. Pauli Girl and then he slapped me on the back and hugged his Sandy again. His eyes sparkled and he told me that Sandy was the one, that all the other girls didn't count and that

Sandy was real and wasn't real love the most wonderful feeling? Sandy beamed at me. She was pretty and petite and very much taken with Jake. The three of us sat together for a long time while Jake bought a lot of beer and insisted that I help him drink it. Sandy put her arms around me and asked me please to help them celebrate their real love and so the day wore on and the party grew bigger and I was soon lost in my corner while all the Green Parrot regulars laughed and hugged and helped Jake and Sandy celebrate.

I don't remember them ever leaving the bar. At some point I found myself walking the highway back to Stock Island and my boat. There weren't many cars on the road that night and it was hot and sticky and I had to go to the bathroom. I waited until I got to the restrooms on the dock where I fumbled with my keys in the dark and stumbled through the door and shut my eyes and leaned against the wall over the urinal. I was still upset about Janie. After I flushed the beer down the pipes, I staggered back to my boat, alone and drunk and drained.

I never knew what became of tiny and petite Sandy. She had such dark eyes and when you saw her looking at Jake, you only thought of kittens and daisies and lady bugs and love. Sometimes you saw the same things in Jake, but there had been so many other ducklings that had followed him into the bars in Key West. With him you never knew. It wasn't long after the Green Parrot celebration that they found Jake lying dead on A1A. He might have known for an instant, driving into the sun that afternoon, that he was about to die. I have often wondered if he thought of Sandy.

CHAPTER TWENTY-TWO

free in the wind...

I didn't feel very well, even after a morning shower. A pot of Bustelo bubbled on the stove but the aroma of the Cuban coffee wasn't doing it for once. I leaned against the counter and shut my eyes and tried to clear the halyards banging in my head.

"Lew, come on out," someone shouted. "It's lunchtime already."

I recognized Frank's squeaky voice.

"Hey, we need some help," he yelled. "We've got a project going."

"Just a second," I said.

"Come on over to *Shady Lady*. I'll make you a Bloody Mary to kill your hangover."

"Sure. Give me a minute to get dressed. I had a little problem last night."

"I hope you like Tabasco sauce."

Frank was a corporate lawyer from Cleveland who probably didn't make the kind of money he wanted to spend. He stood only 5′6″ in his platform shoes, but he enjoyed having the tallest mast in the marina. *Shady Lady* floated like a concrete sewer pipe over near the office. When I stepped aboard, Pirate Jud handed me one of Frank's Tupperware refrigerator containers filled with a Bloody Mary, mostly vodka and hot sauce. Frank's buddy from Cleveland had already started his.

"Holy shit, Frank," Don said. "You're supposed to make a Bloody Mary with some tomato juice."

"Just shut up and drink it," Frank replied. "You won't remember what it was made with in a minute."

I sipped the Bloody Mary and scalded my throat with the vodka. Frank looked back at me.

"We're on our way to put *Shady Lady* in the charter service up in Miami. Don is here to take pictures for the advertising brochures. I was hoping you could take your boat out today so we could get the photos. What do you think?"

"Sure. We can do that. Are you too cheap to buy more tomato juice?"

Frank chuckled and got out the Tabasco sauce for another round. The vodka woke me up and I finished the second Bloody Mary in a hurry and then walked back to my boat. Don followed behind with a canvas bag full of photography gear.

I backed *Sun Flower* out of her slip and motored into the channel where we set about raising the main and genoa. *Shady Lady* puffed her way out of the marina, but it took Frank and Pirate Jud a long time to sort out the lines and sheets and halyards. They didn't get everything full and drawing until the boat was a mile up Hawk Channel. Frank's big stone ketch finally came around on course and moved up the flats looking much better in the mid-afternoon light.

Don snapped several dozen pictures while clinging to the starboard shrouds. When he had enough exposures, I sailed into the lee and luffed the genoa to slow the boat. Don stepped from the deck onto the ladder and climbed aboard *Shady Lady* without dropping his bag of cameras. I let the boat fall away and then sheeted in the genoa and sailed alongside while Frank and Pirate Jud yelled and waved good-bye. I could see Don snapping away with his cameras. I thought *Sun Flower* made a good showing, but I turned to go back down Hawk Channel toward the Sand Key light that stood proud in the distance.

I let *Sun Flower* run free in the wind that was hot from the south. She pranced over the flat water at ten knots until I neared the channel through the reef. I sheeted the genoa home and hardened into the wind and in a moment, the boat was footing toward the pass at twelve knots with the windward hull flying high and the sound of the water roaring through the tunnel.

The channel near Sand Key was open to the deep water beyond the reef. I steered close to the spit of sand and shot through the gap where *Sun Flower* romped into the swells rolling in from the south. The wind blew across my face and through my hair and I breathed the clean salt air and let *Sun Flower* sail herself on a close reach. I couldn't think of any good reason to turn back for the marina. The bottom was still visible in the clear water, but it didn't take very long before I was in the Gulf Stream where all I could see was the indigo blue of the eternity beneath the keel. There were cans of food still stored in the bilge lockers and the water tanks were nearly full and there were moments of anger because of Janie.

I snapped my harness onto the jackline and walked up to stand near the mast. There were whitecaps forming in the distance. I thought about setting a reef in the main to slow the boat, but I didn't care anymore. I stood and held my face to windward where the wind and the salt spray stung my eyes. Tears formed and ran in streaks on my skin while *Sun Flower* flew over the swells. There were times when I thought we should keep on screaming toward the Bahamas that hid just beyond the horizon, keep on screaming toward the banks and away from Janie and Key West, away from the tequila shooters and the rum and the St. Pauli Girls, away from the biggest mistake I ever made up to that point in my life.

I was several miles out into the open sea before I gave up and jibed the boat to run off the wind toward Sand Key. I don't know why I turned around, but *Sun Flower* settled into a broad reach and began surfing to fifteen knots or better. She seemed a lot happier suddenly, maybe because the motion was far less frantic. I settled into the cockpit as well and watched the Sand Key light grow larger in the distance.

We crossed through the pass in the reef where we slowed to ten knots over the white sand of the flats. Key West became a line of tiny rectangles, and I could make out the stainless pipes and valves and tanks of the desalination plant out on Stock Island. I sailed toward the entrance to the basin where the outriggers of the Singleton shrimp fleet came into view and where the desalination plant ghosted by like a giant Erector Set weathering away in the quiet afternoon sun. The marina was deserted. I could see a gaping hole near the office where *Shady Lady* had been tied. The streaks of algae oozed down the empty wall, glistening black in the low light of the afternoon. *Sun Flower* eased into her slip and waited while I made fast the mooring lines.

After I shut the diesel down, I could feel my headache returning in the silence. My face and hair were windblown from the afternoon sail so I sneaked up to the showers and stood for a long time in the hot water. I toweled and dressed and went back to the boat long enough to get my wallet. I still felt queasy from the night before, maybe from Frank's vodka and Tabasco sauce earlier, and I thought it might have been a mistake to come back. There was no point to it with Janie gone. I locked *Sun Flower* and jumped onto the dock and walked across the field of shrimp nets and gillnets to the transit stop around the corner.

Sometimes the Green Parrot escaped the roving bands of tourists. When I walked in, the jukebox wasn't even playing. I sat in the corner and talked baseball with Old Man Ned who had played Single A ball in the Carolina League. Some of the regulars came through and I got caught up with Key West news, but then Big Jake came in with Drew and took a seat just down the bar. They were both wearing T-shirts from their conch house restoration business. All I saw was the cartoon guy in the front with a slobbering tongue hanging out. It made me think of Janie and her boyfriend and the drooping ladder on the DC-3 that took her away. I drank from the bottles of St. Pauli girl and stared at the filthy silk panels of the parachute hanging loose from the ceiling over the bar.

CHAPTER TWENTY-THREE

slap and hiss...

The quiet mornings near the coffee kiosks gave me a lot of time to reflect on how it was that I found myself alone in Key West. I started to think about those dreams, though, the dreams of sailing the lower Caribbean and running off before the trades, the dreams even of finding that old schooner from the Waterway still riding those swells that rolled in all the way from Africa. I didn't want to give any of that up, even without Janie.

A friend of mine on the docks told me about a yacht delivery position that opened up when The Moorings decided to upgrade their fleet in Tortola. I did a telephone interview for the job and was hired to sail some of the new Gulfstars down to the Caribbean to be placed in the charter service. Key West seemed deserted to me so I left the moveable fiestas behind, hoping Janie and the bitter memories would fade away astern.

Most of the time the delivery trips were benign affairs where we sailed from St. Pete down through the Bahamas and out into the Atlantic for as long as we could stand it before turning right and heading for the British Virgin Islands. It was a good time to be alone and free, except for Janie, and I lost myself at sea and tried to put my life back in order. There were days of flat calms and bright skies and days of sailing that convinced you it was only a sport reserved for kings when the green shadows of St. Thomas and Tortola materialized in the distance off the bow. We sailed into Road Harbour where we tied up at The Moorings docks and then headed into Drake's Pub to order fish and chips and pints of Double Diamond bitters. It was always a relief to be done with the sea and to be drawn into the reggae sounds of the pub.

The delivery trips were great and innocent fun and proved to be somewhat cathartic, at least while we were at sea, and I began to ease into a zone that bordered on complacence. I enjoyed the blue water sailing for the pure love of it, but these benign feelings ended abruptly when we were caught offshore when the wind and the rain and the sea were gray and violent, very much like the hurricanes that marched through Myrtle Beach in the fall. On this day there were high ribbons of cloud surging across the sky, riding hard before the winds aloft. We stood on the foredeck to watch the oncoming storm while the leading edge took on the appearance of an oily, sulfurous guillotine. By evening the seas were thirty-five to forty feet and breaking from astern. The boat rode to the top of the swells like a roller coaster in reverse. At the crests of the combers, you could see the spray being driven before the wind in a blinding, liquid blizzard. After the seas roared through, the boat slid down the backside into the calm of the troughs, but then another wall of blue heaved itself astern of us to lift the boat again, forcing it up into the screaming wind once more.

We furled the sails early on and shut the boat up tight. When I took over the helm, I sat behind the binnacle with double harnesses around my chest and safety lanyards hanked to the primary winches on either side of the cockpit. The boat was surfing at hull speed under bare poles. Mountains of blue overtook us from behind and lifted us stern first while we climbed backwards toward the sky for an eternity. At the top of the crests the wind and the rain and the driving spray blasted away at my back and then we dropped down the chutes to the calm of the troughs below, and then up again to the crests where the wind screamed and then down and down once more into the abyss, boiling into the troughs like the Bowery Boys in an endless carnival ride from hell. It was hard to keep the boat from broaching, but I felt an odd sensation of nerves and of power and control. I remember thinking in all of that chaos it was the greatest sailing I had ever done.

About eight o'clock that evening, Milton stepped out of the companionway to relieve me at the helm. After we had secured him with the double harnesses, I went below to lie down on the port settee to get some sleep. I hadn't been there very long before I was thrown from the settee onto the cabin ceiling where I was pinned under a ton of water that had burst through the companionway hatch. Dave and I were trapped against the overhead, but the boat snapped upright and we were thrown again, this time onto the cabin floor. Both of us were tossed about like rag dolls in a laundromat, choking on the salt water that surged through the salon. When I could stand again, I rushed to the open companionway to see about Milton who was strapped in at the helm. The binnacle had stayed attached to the cockpit sole and I could see him hanging onto the wheel and staring back through the dark at the next waves tumbling toward us from astern. It hurt him to breathe and his voice didn't have a lot of volume and I thought he might have broken some ribs. He wanted to stay at the helm even though you could see the imprint where his chest had bent the stainless steering wheel.

The breaking seas continued to run huge through the night. Milton kept the boat running before the wind and the walls of black water while Dave and I stayed below to put the interior of the boat together. We found dishes from the lockers over the galley tossed into the forward cabin. The companionway ladder had torn loose from its mount and had bounced around the main salon ripping holes in the headliner and gouging the finish of the teak bulkheads. My sextant had come adrift from its locker and I couldn't find it until Dave lifted the settee cushion and looked into the bins below and found it lodged beneath several dozen cans of vegetables and a five-pound bag of oranges.

At midnight when it came time for Dave's turn at the helm, we struggled topside to help Milton out of the harnesses. There was no stern rail left at the back of the boat. When I looked at the rig, the masthead light was still burning but the VHF antenna hung loose, banging around on its coax fittings. I don't know why the mast hadn't gone overboard. The rig looked fine in the dark and I thought it would stay if we weren't rolled over again.

When we climbed once again one of the forty footers that boiled through, I caught a glimpse of the lights from a freighter not far away. After we secured Dave at the helm, I went below to call them on the VHF to check our position and to let them know who we were in case of further trouble. When the radio operator answered, he said they had seen our masthead light but it disappeared from view when we dropped into the troughs. I told him our light was fifty feet over the water and he only just whistled into his mike. He then told me the ship's anemometer had recorded winds of seventy knots and gusts to eighty-five and they lost deck cargo overboard when a container had come adrift in the heavy seas. There was nothing for them to do but continue on their way. It was good to hear the man's voice and it was some small relief that our radio still worked even with an antenna that bounced around at the end of the coax cable with each passing swell. I thanked the radio operator for his help and when I signed off, I sagged against the bulkhead, exhausted and spent and hollow again inside for what the rest of the night might bring.

By daybreak the wind had lightened to no more than a morning breeze. The seas were huge but only rolling through instead of breaking. The three of us sat in the cockpit nursing our cuts and bruises and sipping coffee fresh from the stove that still lit. All we could smell was diesel fuel that had spilled from the spare containers when the stern rail had torn free. There was a fine brown film over the decks that made them slippery and treacherous and it covered the cockpit where we sat. The coffee was good and the sunrise was bright and we laughed about our narrow escape, but the laughter was hollow and the pain was real and so we mostly sat in silence.

Milton could only tell us in his wounded voice that he had kept the boat dead before the wind. Off to starboard a rogue wave had towered out of nowhere to slam into us from the side and there was nothing he could do. We were thrown onto our beam ends and that's when the companionway hatch blew open to flood the interior. He had been smashed into the wheel where he hung on to keep from going overboard. There was no doubt the double safety harnesses had saved his life. Had we lost him in the capsize, I don't know that Dave and I would ever have recovered.

Kenting Tropical Forest Park is on the southwest corner of Taiwan. When you drive past the entrance, you come to a high bluff near the Oluan Pi light that stares across the South China Sea and the Straits of Taiwan to the mainland. On top of the bluff there is a bunker and a gun emplacement that looks like a John Wayne movie set, but there are real soldiers hanging around who are serious about what might come across the sea. They stand their positions and look smartly about and you get a vague feeling someone is winding the strings too tightly on an expensive guitar.

All the tourists who come to visit the park stop to view the bunker and the soldiers. They speak in hushed Mandarin or in the hushed street dialects of Taiwan and they marvel over the beauty of the park and the view to the sea that stretches away to the enemy. While the tourists are there, the soldiers snap this way and that and everyone gets a good view of the side arms they carry at their waists. The big artillery pieces that point toward mainland China are locked and loaded and waiting. If you stay too long, the soldiers will ask you to leave and so the view of the sea far below is only fleeting. In the rush to move you forget the tension for a moment, but once in the car you realize that the bunker on the bluff is not a movie set. It is no place to sit and do nothing.

On my first visit to the bunker I got the odd notion I was standing too near a set of rusted-out garage door springs. It was the same feeling I got being at sea after dark with the wind screaming and the seas walling up from astern. I didn't wait for the serious soldiers to ask me to move along. I drove back to the park and down to the beach far below where I left my car and wandered down to the water. There was no surf then, only a six-inch shore break that lapped at the sand and slid up the beach in an endless cycle of slap and hiss, slap and hiss. I walked alone toward the headland that loomed in the distance. The soldiers were up there waiting with their hands on their hips and their cannons aimed at the Chinese hordes that were sure to arrive any minute, but here there was only the slap and the hiss of the water and the sounds of the sea birds aloft.

With the heat of the afternoon sun sprawling over the beach, it turned out to be the kind of day where you want to sit and do nothing. It was a relief to be alone when there was no tension in the air. There were no gusts of wind or driving rain and no booming of a frenzied surf line, no halyards banging time against a hollow mast, no backwards roller coaster rides to churn your stomach in the middle of the night, and no armed guards who only glare at you out of the cold. I flopped on the warm sand and stared out to sea toward mainland China. I sat there by myself thinking about the capsize and about those summer days in Myrtle

Beach, thinking about what had gone wrong in Key West. There was only a quiet breeze, soft like petals of down, and all I could hear were the birds in the distance and the gentle sound of the sea, slap and hiss, slap and hiss.

CHAPTER TWENTY-FOUR

a bottomless pool...

I made several of the trips from Florida to Tortola in the British Virgin Islands delivering the new Gulfstar 37's that were put into the charter trade. I spent countless days at sea simply watching the trade wind clouds and the horizon and the sea itself, trackless and endless and blue as the twilight sky. There were times when I didn't want to see Tortola rising green in the distance, but would rather have stayed at sea lost in the ultimate version of freedom.

Because we spent a good part of the time sailing in the temperate zone north of the trades, the wind was never reliable. There were days of glorious sailing when the Northers followed us out to sea, but there were always days of dead calm. I didn't much care to start the diesel too soon, preferring instead to flop about for a bit to wait for the first of the zephyrs that would signal the oncoming winds.

On one of those quiet afternoons, I sneaked below to make a cup of instant coffee. When I got back out to the cockpit, I sat alone and glanced about at the sea that looked like stretched gray silk it was so calm. I sat there for a long time sipping the coffee and watching for sharks and big fish, but you couldn't tell where the sea ended and the sky began. It seemed suddenly like the boat was the earth and we were ghosting along at the center of a silver-gray universe.

When the fitful breeze fell away to nothing, the boat began to wallow and drift. The air was so breathless, the sound of the mast and boom clanging about reminded me of a St. Pete cafeteria. The noise brought Erik and Poupon outside. The three of us sat in the cockpit watching for signs of some wind, but there was nothing but the silver-gray glass of the surface and the tropical heat that hung limp and lethargic and smothering. We had no Bimini cover because the boat was new so the heat soon became unbearable in the cockpit. We decided to stream a line astern and dive into the sea to cool off.

Poupon was first into the water. I followed close behind while Erik stayed aboard to watch in case the wind began to fill. The air was absolutely still and so the boat drifted in lazy, random circles while we cast free of the line and dived and splashed like kids in a neighborhood pool. After ten minutes or so, I swam back to the ladder to give Erik some time in the water. Before I could lift myself aboard, though, a breath like a feather out of nowhere caught us from behind and the boat slipped away. I grabbed the trailing line but Poupon was too far astern. We left him behind treading water. After I scrambled aboard, Erik and I started the engine to bring the boat around. Poupon's head was a bare speck on the sea when we turned and it was hard keeping him in view. I remember thinking how stupid it was for us to let him get so far behind. There were times after we got underway that we couldn't see Poupon at all and I was nervous that we had lost him. We sailed a reciprocal course and found him soon enough, treading water on his back and singing a French bar song. You would have thought him to be swimming off the beach in Biarritz instead of floating in mid-Atlantic 300 miles north of the Virgin Islands. We pulled him aboard and then broke out a bottle of rum and sat together in the cockpit where it was great fun to laugh, but there were moments when I knew how close we had come to death.

A friend of mine told me a story once how he had been at the helm of a boat he was delivering when the night wind dropped to nothing. The heat made him sweat and in a moment of stupidity, he decided to jump overboard with a line in his hand. The crew had been doing this earlier in the day when the wind had been fitful. Bill only meant to cool off for a moment and then swing back aboard. To his horror the line he had grabbed wasn't made fast. When the boat drifted away, the line came loose in his hands and he was alone in mid-ocean. He swam for his life but he couldn't catch the boat. He screamed and choked in salt water and tried to wake the sleeping crew, but there was only silence aboard while the boat drifted into the dark of the night. Bill was left alone in the sea facing a certain death.

Some hours later one of the guys stepped up the companionway ladder to relieve Bill at the helm, only to discover an empty cockpit instead. He woke the other crew member and together they calculated the speed and the drift of the boat and turned about to sail a reciprocal course. They were careful to keep a lookout on the mast spreaders to watch, but as the hours wore on toward late afternoon the next day, neither of them thought Bill could have survived. It was nearly sundown when the lookout shouted about something in the water. When the guy at the helm turned the boat to see, they were astonished to find Bill, barely conscious and slipping beneath the surface with each swell that rolled through.

Years after this had happened I listened to Bill tell me his story. I could see him wince at the memory. I thought again of Poupon and how fast he had been left astern and how small he became in the big ocean and how devastating it would have been if we hadn't been able to find him. I don't know if Poupon ever understood this, but with Bill you could see in his face the shadows as they passed.

On the day *Sun Flower* crossed over the Puerto Rican Trench, the trade winds died away to nothing. We wallowed in the heat and the stillness until Buddy threw a line into the water and dived overboard to see what it was like to swim in an ocean over five miles deep. I grabbed a mask and some fins and followed a moment later. When I cleared my mask, I was astounded at how far I could see. I took a big breath and dived deep into the abyss where I stopped and hung motionless in the incredible depth. There were no sounds down there and the water temperature nearly matched my body heat. It felt like being in the womb again.

I jumped into a pool when I was two and when the bubbles cleared, I hung motionless near the bottom, peering into the same kind of gloom. I remember rolling about, drifting more than anything, and I kept looking through the gloom and at the sunlight above that sparkled bright on the surface. Suddenly a hand crashed through the sunlight and groped for me and pulled me up and out of the pool. I was set hard upon the concrete where my mother fussed over me and dried my face and my arms and legs, but I could feel the cold wind blowing across the pool deck.

I glanced at the hulls of *Sun Flower* high above. An ominous shadow hung around that I couldn't shake, like that cold wind when I was two, and I was suddenly afraid of the awesome depth beneath me. I kicked for the surface where I burst into the sunlight and took huge gulps of the warm air and then floated on my back, wondering why the deep water made me so afraid. I swam over to the trailing line and pulled myself toward the ladder. I wanted to be back in the cockpit again where I would be free of the chill sent up from the depths nearly thirty thousand feet down.

It is still and quiet outside the café right now. The humidity has settled in low to the ground, held in check by a heavy Southern heat that feels like wet woolen blankets. The water has that silky look once more and it makes you feel like you are at the center of the universe again. Weather like this reminds you of what it was like before you were born. Jumping into eighty-three degree water so deep you can't comprehend its depth because you can't comprehend eternity seems very much like jumping into the warm eternity of the womb. With Poupon treading water so far behind and the miracle that Bill was ever found, there are moments when I wonder what it might have been like for them had they not been recovered, hanging motionless in a bottomless pool from a long time ago, slipping away into eternity like stillborn children. I still shiver at those shadows.

/) CHAPTER TWENTY-FIVE /)

Willy Dog...

Yesterday I was sitting by the window in the front of the Starbucks when I noticed a new Mustang fresh from the dealer pulling into a parking spot across the way. I liked the color of the car and I was staring at it when I noticed it had been parked next to a broken and rusted Toyota pick-up. The truck looked bow-down to me, like the front springs were shot, and a lot of dents and prangs rumpled the body. Someone had sprayed a good part of the cab with cheap red primer and hadn't bothered to tape the windows. The Toyota is still there today. The front tires are nearly flat and the windshield is broken and the rearview mirrors have been taken off and there is no license plate in the frame. The truck has probably has been left for dead. One of the kids in the coffeehouse will have to call the police to have it towed away.

Someone parked a late-model van in front of the conch house where Franco lived in Key West. There were no license plates on it, but it was new and clean and everyone thought at first one of the neighbors owned it and so it stayed in the street in front of Franco's for a long time. Willy Dog took to peeing on the wheels and on the sides and the dust that collected on the roof became streaks down the windows every time it rained. People would stop on the sidewalk to peer into the interior, but still it sat neglected.

Franco decided the van might have belonged to a drug runner and that maybe it had been abandoned after a deal had gone down. This idea scared some of the locals and they wouldn't go near it again because of their fear of the dope runners. The police never ticketed the van because it was still new and since everyone was afraid of who might own the van, no one called to complain. The van sat deteriorating in the sun and didn't get hauled away. Finally someone came in the middle of the night and stole the wheels. It sagged on its axles like an old workboat that had sprung a leak and had settled into the mud. The van only got dirtier and where Willy Dog had peed, you could see rust forming in patterns that looked like the rocker panels had been sprayed with birdshot.

In the following weeks someone sneaked through and removed the bumpers and the two front doors. Franco told me he had heard clanking noises in the middle of the night, but he had only just met this girl and already they were in bed so he didn't bother to get up to see. When the seats turned up missing and the paperboy broke the windshield with a softball bat, Franco finally called the police. After they arrived, a crowd of locals showed up to watch the proceedings. The police informed Franco that he was to be charged the towing fee. Franco said he wasn't going to pay shit because he didn't own the van. It took the police a lot of radio calls and a lot of talking in the middle of the street to figure out the van had been stolen in Miami.

The sun was going down and most of the crowd had gone off to Mallory Square when the carcass of the van was winched onto a flatbed truck and hauled to an impound yard. Franco was glad to see it go. I didn't want to be angry about Janie any longer. After the van disappeared around the corner, I bought Franco a St. Pauli Girl at the Green Parrot Bar and Sub Shop, but not before Willy Dog had peed on the front wheel of the police car and wandered off down the street.

CHAPTER TWENTY-SIX

Big Max...

I was on my way up to the shower one day when Margie and the gang came running up and hugged my legs and begged me to let them use the fishing gear. I dragged a few kids along the dock, but they wouldn't let go. We turned around like some kind of eight-legged paramecium and staggered back to *Sun Flower* to dig the fishing rods out of the starboard hull. I showed them how to open the bail on the spinners and how to keep the line level on the little Penn, but they were too excited to listen. They all ran off toward the bar to see if the fry-cook had any shrimp. I walked up the dock again and headed for the showers. I could hear the gang over on the wall behind *Cimarron*, laughing and trying to figure out the spinners. I kept trudging on my way with my head down hoping they wouldn't see. All I wanted to do was stand in some cool water and let it soak away my hangover. When I got back to *Sun Flower*, I was bowled over by the heat below and by the pile of filthy clothes on the cabin sole. I put all of them, including the deck shoes, into a plastic bag and threw them into the cockpit and then slid the companionway door shut. I walked over to the bar for some coffee.

"Your friends were in here panhandling for bait," the bartender said.

"Did they do any good?"

"Slow Time gave them a baggy full of sixteen/twenties. They all ran down to the end of the dock to the deep water. They might catch something."

"Got any coffee?"

"Comin' up."

I sat at one of the bar stools and drank coffee so thin I could see through it to the bottom of the cup. When Slow Time came over to see if I wanted more, I nodded toward the kitchen.

"Any bacon and eggs left?"

"I'll bring it over."

I moved to one of the vinyl booths where I could look out across the marina through an open window. I couldn't see the end of the dock, but *Sun Flower* was there along with *Cimarron* and the black schooner *Tonga,* and several of the houseboats in various states of disrepair. The two mackerel boats from the party the night before were dark and silent.

After breakfast I walked out to the end of the dock where I found Margie holding a spinner upright and reeling backwards next to Joseph who had the Penn. There were five half-pound snappers in the bucket. The kids squealed with delight over the fish and wanted me to stay and watch. I sat on the crumbling cement with my feet over the side while they turned serious again and tossed their lines ten or twelve feet out where the marl dropped away to the deep water of the channel. The heat of the afternoon got me sweating soon enough, but there was some breeze blowing in from the southwest that rippled the waters and made the fishing lines billow in long, lazy arcs.

I could see from the end of the dock clear beyond the channel entrance to the reef some three miles out. The thin water of the flats glowed pale green in the high afternoon sun. I stared at the green sea out toward the reef. It didn't feel much different than sitting on the banks of the Intracoastal Waterway near the bridge on Highway 501 and watching the tugs go by when I was a kid. I leaned back on my arms and sat with my legs hanging over the side while Margie and Joseph jabbered away.

One of the kids screamed. I looked over to see Joseph's rod bent double and the drag on the Penn was screaming at the same pitch. I jumped to my feet to lend a hand but there was a loud snap. Joseph fell on the seat of his pants while the line flew back and hung limp at the end of the rod. Margie and Little Bill were jumping up and down and laughing at poor Joseph who was near tears at the giant fish that got away. I helped him up, but the fish had stolen most of the line from the reel along with the hook and sinker. Margie ran up the dock to tell Pirate Jud while Little Bill went back to work trying to land the giant fish himself. I tried to console poor Joseph who had nothing to show for his big adventure.

"You must have hooked Big Max," I said.

"Big Max?"

"Yeah...The biggest barracuda in the channel."

"Have you seen him?"

"Yep, right here in the basin when I was diving."

Joseph's eyes got huge and he looked up at me and then he took off after Margie yelling all the way up the dock trying to tell her about Big Max. Little Bill kept fishing. I flopped back on the concrete and dangled my feet over the edge again hoping Big Max would come back.

Little Bill caught another snapper, reeling the spinner wrong side up and backward, but he landed the fish and let it flop around in the bucket. It had swallowed the hook and I pried at the shank before it popped free and then there were six snappers for dinner. Poor Margie hadn't caught any of them. When she came back, she plopped herself down and looked in the bucket. I reeled in her line and found an empty hook at the end so I took the last shrimp and threaded it in place and cast the spinner far out into the channel. I told her to hang onto the rod because Big Max was still lurking around out there. I sat down beside her and stared at the deep water sparkling in the afternoon sun and waited. Joseph came back, but there were no more shrimp so he sat on the concrete and watched while Margie tried to put the English on Big Max. Far down the channel a shrimper headed in for the Singleton docks and I watched it while the kids chattered away. Margie didn't move a muscle, but Joseph jumped to his feet.

"You got one...You got one," he shouted.

Margie stood up to reel backwards again. I turned the rod over for her since it was set up for left-hand retrieve, but she didn't like that and reeled backwards with her right again while the line stretched far into the channel. There was so much line out it took a long time for her to land the fish. After it flopped onto the dock, Margie had a fat, two-pound snapper that got the kids jabbering again.

They carried the gear back to *Sun Flower* where I stowed it away. I filleted the snappers and put the fish in separate baggies. The kids thanked me for the help and then ran off laughing about Joseph landing on his britches. I stepped below to try to clean the smell of shrimp and snapper from my hands with dish soap and then I sat down at the dinette thinking about Big Max and wondering where Janie had gone.

The kids had cured my hangover for most of the afternoon, but in the quiet of the cabin I could feel some queasiness returning. I flopped on the settee to shut my eyes for a minute. The cabin heat made me sweat again and I caught myself drooling and I sat up. I had been asleep for nearly an hour. The light from beyond the open companionway turned red from the sunset and I climbed into the cockpit and watched a high trace of cloud change color while I tried to work the kinks out of my neck.

I yawned again and stepped off the boat and made my way out to the end of the dock where I sat down and looked out toward the flats, pink in the setting sun. It felt like I was on the river again. I wondered if Margie and Little Bill and Joseph would look back and think about shrimpers and sailing boats and pale green water that stretched all the way to infinity. I leaned back on my arms and sat with my legs hanging while the breeze drifted through. The air was laced with salt and with the smell of seaweed and it wasn't as sweet or as heavy as the air in the cypress swamps, but there wasn't much difference. A world had gone by but I could have been sitting on the Intracoastal Waterway in the Lowcountry of South Carolina. There were shrimpers outside in the channel and I could see a day charter on its way back from the reef. I watched the boats until they disappeared. I rubbed my eyes and looked beyond the flats toward the Gulf Stream in the fading light. I was thinking about that schooner again, wondering where it headed after it disappeared through the drawbridge.

CHAPTER TWENTY-SEVEN

Ben Gunn...

When someone opened the door, a gust of wind swept an older couple through the entry with waving scarves and billowing overcoats. They blustered about and blew on their hands and made a fuss over each other and finally took a seat near me. The lady left and returned with a *palmier* on a plate and two mugs, but the man complained to his wife that his latté was not hot enough so she has taken his mug back. While the latté was being redone, the old man sat alone and munched on his palmier by breaking bits of it off. When he chewed, it looked as if he were using only his front teeth and it gave him the appearance of a stage actor in an English comedy. He was thin and bony and quick and his hands darted from palmier to mouth to napkin to chin and he looked for his wife with glances that reminded you of Wimbledon. Watching him I found myself thinking of "Waiting for God" on television and dime novels with characters that hang around the waterfront and accents that sound like pound notes fresh from the mint.

There is a narrow and winding cut through the coral into the inner harbor of Chub Cay in the Bahamas. If you make your way through the pass to the docks, an official will come down to your boat and stamp your papers and clear you through customs. The island is quiet and there is no parading of uniforms and badges and no blustering and you can feel in the air the kind of out-island ambience that makes you want to move slowly into the day, alone and free.

It's like Nassau didn't exist.

Chub Cay is expensive, though, and the locals cater to the wealthy sport fishing crowd from Miami. The marina is filled with boats made of fiberglass and stainless and teak veneer and they all have names like *Wet Dream* and *Kamonawanalaya*. Not very often do you find a boat there that's meant for the real sea, nor do you find the kind of character who sits across from me in the coffeehouse making great work out of a simple palmier.

There was once, though, a thin and wiry old man who came puttering through the cut in the coral driving a surplus U. S. Navy landing craft. He tied up at the marina next to the gold-platers from Miami and waited for the customs man. I was on island time then and so I left my boat and walked over to the bar and bought a rum and tonic and then I went to see the old man who was hopping around his Mike-boat like Ben Gunn. He smiled at me and then continued doing whatever it was he found so urgent. I looked at the landing craft and watched him dart here and there. So much movement was going on that it made me think of a friend of mine in Key West who spent so much time bouncing from project to project that nothing ever got done and then his girlfriend, who was the heiress to a huge cookie fortune, left him for a good and quiet kid from Carlinville, Illinois.

I finally asked this old guy what he was doing in Chub Cay with a Mike-boat. In a quick Billy Barty voice, he told me he was delivering it to the British Virgin Islands for some new owners. The trade winds, though, even in late spring, can be fierce. Heading south and east puts the wind and the current right on the nose. In a flat-bowed landing craft, you might as well be driving a snow plow. The old man was serious and full of confidence.

"Do you have any charts of Haiti and the Dominican Republic?" he asked. "I want to island-hop all the way down."

"I never bought any," I said. "Too many stories of disappearing dinghies. I had friends who were boarded at night off Haiti by a band of armed men. I don't want to put myself in that kind of situation."

The little man kept hopping about and it was apparent he hadn't heard anything I said.

"How about a rum and tonic?" I asked. "We can look around the marina for some charts."

"Thanks," he said, "but I have too much to do."

The next morning when I went back to the bar for coffee, I saw that the old man and the Mike-boat had disappeared. The bartender didn't see him leave. When I left to return to the dock, I stood next to the empty slot where the landing craft had been tied. I could see from there down the cut toward the banks beyond Chub Cay. I thought about the old man slamming into the steep head seas built up by the wind and the current and I hoped he had stopped in Nassau or gone down the Exumas to George Town. I hoped he had taken the time to stop in a bar for rum and tonic.

Maybe someone talked to the old man and maybe he had changed his mind and decided to stay for a while. Maybe the engines in the Mike-boat quit and he had to abandon his journey somewhere down islands in the Turks and Caicos. In later months when I sailed to the Virgin Islands, I looked for the old man and the landing craft, but he wasn't there. No one I knew had ever heard of him.

Jim Brown, the trimaran designer, wrote a very good construction manual in which he kept referring to the sea as an "absolute" element. No one who goes offshore ever thinks of it in any other terms, so you take what the sea gives and accept responsibility for your failings and work to reach your destination. Sometimes those failings are minimal and the Adventure is grand, but sometimes you can stack your own deck and doom yourself to failure. I don't know if the old man and his landing craft were in that category, but Bimini bread and conch chowder with hot sherry are national delicacies in the Bahamas. There are pubs that serve John Courage and Double Diamond in big-handled pints and the rum and tonic and lime is brisk and fizzy and very British. The ocean water down there is clear and the winds from the east blow fresh across the banks, and the sun is always warm on the coarse sand of the islands. It's nice to think he never left.

CHAPTER TWENTY-EIGHT

field of flowers...

There aren't many people coming into the Uphill Grind today, but a young lady walked in a minute ago wearing a Dinosaur National Monument sweatshirt. I wouldn't have noticed except she knows the kid behind the counter. They were laughing about something that happened the night before. I didn't see her sweatshirt until she turned to smile at me, but then we talked a bit about the dinosaurs and about the Green River and about the beautiful farm downstream from the cliffs. After she left, I kept thinking about the tumble of fossils preserved in the main building in the monument up in Utah.

I spent a day on a geology field trip once helping to excavate the fossil remains of a baleen whale from high on a hill in the Hunter-Liggett Military Reservation in Monterey County. The ten-foot whale dated from the middle-Miocene and was a rare find in those parts even though the Monterey shale is loaded in places with leaves and crabs and other small fossil organisms.

Mr. Bristow was the geology professor at Monterey Peninsula College and the greatest teacher I ever had. He made sure we didn't damage the whale in any way and so there we were at the top of that hill watching the advanced students dig out a 15-million-year-old fossil skeleton with toothpicks, dental tools, and paint brushes in mid-ninety-degree heat. I wasn't sure what the objectives were other than some lessons in the discipline involved in Paleontology and the recovery of the whale bones for display at the school, but the bones looked to me like a modern whale had flopped up the hill and died on the spot. Whales are such social creatures. It made me wonder if another skeleton was hidden in the shale somewhere near.

South of here in Mexican waters, a whale showed up one evening and followed *Sun Flower* about one hundred yards astern. Whales can be bold and unafraid, but this one was timid and wouldn't come much closer. For two days it cruised with us, sometimes swimming abreast of us and sometimes pulling slightly ahead, but nearly always falling back again to trail us at a safe distance. At night we lost sight of it, but you could hear the whale in the dark blowing its lungs free of air. It sounded like the brakes on a city bus. In the morning when it played near the surface, you could see the sun dancing from the black glycerin of its back. We thought at first it might have been a finback, but they are so rare that in the end we decided it was a lonely humpback looking for company. We never tired of watching it. For those two days the whale lolled about in the sun and cruised south with us and we were content just knowing it was there. On the third day when the sun came up, we looked astern to see our friend, but the whale had drifted away in the night. We didn't see it again.

Sailing in the lee of Dominica in the Caribbean, I was alone in the cockpit of *Sun Flower* watching the island recede in the distance. We were only ghosting then and the water was flat. There were no sounds and it was nice to sit there and look at the trickling wake of the boat in the water and at the green of the hills fading away. Out of nowhere a whale exploded from the sea to rise two-thirds of its body length out of the water. It was so close you could see spilling from its mouth the foam and the bodies of the krill. The whale crashed to the surface and there was the white of the foam and the red of the krill and the reverberant thunder of the impact, and then all of it died away and I was ghosting along in the cockpit of *Sun Flower* once again, looking at the green of the island in the distance.

We sailed one of the Gulfstar 37's far out into the Atlantic on a day when there was a ten-foot following sea and a twenty-knot wind from the port quarter and not a cloud in the sky. Astern of us a whale began surfing like a dolphin in the swells. We watched it for hours while it gradually worked its way closer until it came abreast of us where it rolled on its side and drifted to within a few feet of the boat. One of the guys brought up the old joke about it being in love, but the whale was so close I wasn't sure there was any joke at all. I hung onto the starboard shrouds and watched the whale. When it rolled on its side, it looked at me through an eye that was only a foot beneath the surface.

All of us were a little apprehensive because the whale was longer than the Gulfstar, but it only surfed along with us and rolled on its side to watch. We got used to having the company and soon we were talking and waving to it like it was just a giant puppy. Our whale stayed with us for the whole afternoon, but during the evening hours it fell astern of us again and then swam away to the south. We were sad to see it go.

When you first see the island of Pico, it stands proud against the sky like a giant brown pylon rising out of the sea in the Azores. There are clouds tearing away from the peak and streaming in the wind like an eternal gray eruption. The mountain itself is stark and cold and beautiful. We rode the ferry across the channel from Horta and hired a driver who took us on a road trip around the base of the volcano that forms the backbone of the island. We sat in the back of the truck and drink from a five-liter jug the deep red wine that stains your teeth and warms your stomach.

There are no fences on Pico, only rock walls built of basalt from the old lava flows and hedgerows of blue hydrangeas in full bloom that stretch for miles. We drank the red wine and stained our teeth purple and watched the old-world landscape rushing by from the back of the truck. The island seemed like an endless green checkerboard with the squares outlined in blue.

Late in the afternoon the driver left the main road and bounced us in his truck down a rutted asphalt path that led to the flats near the sea. There were some buildings down there and some docks and a crane. A great clump of pale entrails floated in the water. Across the road a field of flowers and grass fell away toward the cliffs, littered with giant bones bleaching in the sun, some already white and some with blackened flesh that had not yet been picked away by the scavengers. The stench of death drifted like swamp gas in the air.

The driver waited while we walked among the bones and the rotting flesh. When I looked at the mass of white floating in the water near the landing, I backed from the docks and turned away. We toured the buildings then where there were great cauldrons and iron gates and massive pipes and valves and steel scoop nets and brick ovens that looked like Auschwitz. Standing near them you couldn't say much and so you only looked to see where some of the machinery was made and how old it was and you made small talk with your friends and then you walked out into the air to breathe again the stench from the field of rotting flesh.

The islanders didn't kill very many. An international agreement allowed them to harvest whales if they harpooned them by hand from boats powered only by oars and so they kept lookouts stationed around the island. When whales were spotted, they started the diesels in the big powerboats and towed the whaleboats full of oarsmen out to sea where they were cast adrift near the pods and where the men in the boats would drive their harpoons and lances into the whales that are such social animals. The whales that drowned in their own blood were towed by the powerboats back to the landing and to the sharpened flensing knives that lay on the dock and to the black iron cauldrons that seethed in the buildings, and to the flowers that waited in the big field of bones.

When we paid the driver his fee and then climbed aboard the ferry for the ride back to Horta, someone in the crowd broke out a bottle of Angelique and offered it to the strangers. I took the first swallow and passed it along to a lady friend while the liquor burned its way down. It tasted good in the cold North Atlantic wind that blew down the channel. The ferry wasn't big and it bounced and rolled in the chop, but you could see to the north and to the south the open sea. In between the gulps of Angelique, I looked to the horizon beyond the islands. There were no whales to see.

The baleen whale from the hill in Monterey County used to be displayed in a window box on the junior college campus, but I haven't been back to visit the school in years. I don't know if the fossil skeleton is still there. I spend a lot of time on the beaches nearby and in the winter when the California grays are migrating through you can see the whales spouting along the horizon and sometimes close in, just beyond the granite boulders that line the shore. There are thousands of the whales now since they are no longer hunted. They migrate south to Mexico to do their mating and calving and then they return to the cold North Pacific in early spring. You can see them just by staring out to sea from the rocks in Monterey.

The California grays haven't changed much from the middle-Miocene. If you could see the little whale on display at the local community college, you would think it was just the skeleton of a baby gray washed up on the beach. There is a jumble of whale bones bleaching in the sun out in the Azores, but in some millions of years when they have become fossilized, they won't look much like the calf Mr. Bristow found in the hills of southern Monterey County, or like the fossils that lay exposed up in Dinosaur National Monument. All of those animals died of natural causes. The fossils-to-be on the island of Pico have been nicked and cut with the blades of the flensing knives after the flesh was torn and cut away and then rendered in those cauldrons, and the skeletons were wrenched and cut and torn apart and then all the bones that remained were discarded in that big field of flowers waiting across from the dock. Whales are such social creatures, you know, and they'll swim close by just to see. Out in the Azores, the whales that came near never knew what was hidden in those longboats.

/) CHAPTER TWENTY-NINE /)

another kind of Paradise...

I glanced over the shoulder of a young lady as I walked into the Starbucks. She was studying the anatomy of a generic fish, the inner workings of which were printed in living color on transparent plastic in the text she was reading. She didn't look up at me and when I found a table, I only glanced at her before taking a seat. Outside beneath the eaves, there are kids clowning around and foot traffic coming through from the shopping center, but it's quiet in here where the girl is studying. She is so involved with her book she probably wouldn't understand there are people in life who aren't capable of sitting in a coffeehouse without closing the textbooks and staring, instead, at the blue of the sky through the windows and thinking of the posters that hang on the walls of the travel agency next door.

Things weren't much different a long time ago when my friends and I struggled through classes while we attended Monterey Peninsula Junior College. There was some confusion over the war in Southeast Asia and the possibility of being forced to go kept a few of us honest enough to stay abreast of our assignments, but there weren't many lofty goals being discussed in the student union. We only studied the minimum to keep our military II-S deferments in place and then we spent hours in the student union drinking circus-water coffee and staring out of the windows, wondering what it would be like to bounce across the Australian Outback in an eighty-eight inch Land Rover. None of us ever made the connection between a college education and potential income and so we moved into adulthood about a half-step removed from the norm, sort of a post-beatnik, pre-hippie generation of wandering souls who never got past staring through the windows while everyone else left us behind. The college kids I knew back then never turned into doctors or lawyers or professionals, just teachers and mailmen and technical writers. Even after all these years we think only of places where we are not, places we can't afford to go except on those rare occasions when we can take on a little more credit card debt.

Staring out of windows and pausing to look at travel posters when you have no money, though, is just another kind of water torture, one that turns into quiet agony as you age and you realize it's probably too late to go bouncing across the Outback. It doesn't matter that we spent most of our adult lives out there somewhere. For my circle of friends, life will forever be another window or another travel poster hanging in the mall that grabs our attention and holds us for a moment while we wonder again what it would be like to be there instead of here.

The Leeward Islands of the Caribbean follow a lazy, stepping-stone curve that bends toward the south to the Windwards where the east-facing island coasts are battered by twenty-knot trades and heavy swells that roll unchecked all the way across the Atlantic. When the trades sweep in from the southeast, they put you hard on the wind in the channels between the islands. *Sun Flower* was in her element. She charged to windward in the heavy air, driving her SumLog to twelve knots or better and clipping the occasional wave top that left her yellow decks sparkling with spray nearly the same temperature as the air. The distance between the islands at that pace was irrelevant. We spent a great many afternoons boiling over the Caribbean, heading for the next green jewel that loomed in the distance.

From Montserrat southeast to Guadelupe and Dominica and on to Martinique, and south to St. Lucia, St. Vincent and the Grenadines, and then south-southwest all the way to Grenada, *Sun Flower* pranced her way up and over those Atlantic swells, streaming white water astern and throwing spray and miniature rainbows into the wind.

On the Caribbean side of the stepping stones, the whistling trades are blocked by the island peaks that sometimes tower 4,000 feet or better. After spending an afternoon blasting to windward, *Sun Flower* sailed into the lee of a high island and was reduced to drifting about in the quiet of the wind shadow. The seas lay flat and the winds faded to whispers and the Caribbean itself looked soft and serene in the low afternoon light. You realized then how tired you were because your eyes were burning and your skin was sticky with salt and all you really wanted to do was drop the hook in the sheltered bay and just sit.

After one of those windward romps from the Iles des Saintes, we sailed into the open roadstead off the tiny village of Portsmouth on the island of Dominica and let the anchor slip over the side. The four of us stood on the cabin roof and looked toward the village asleep in the afternoon sun. You could hear the rustling of the trees in the wind and the whisper of the shore break over the sand, and it was like hearing a voiceless echo from all around. I remembered standing still a block from the beach in Puerto Rico when I was little just so I could hear the rush of the sea. I could never tell its direction, only that the sound was all around, and it seemed so far away. The air in the bay below the trees in Dominica was thick and wet to breathe when it was quiet, but then a gentle rush of the trades would return, cool against your arms, and you turned your face but still there was no direction to the wind, only the softness and the shadows and the silence.

I didn't have much desire to go ashore other than to

get fresh bananas or to poke around the shops, but we rowed over to the small wharf and tied the dinghy to a coconut palm and set out on a short walk to take a break from the boat. The road we were on climbed into the hills east of the town and followed a tiny stream that tumbled clear and cold to the sea. My friends and I didn't talk much, choosing instead to listen to the wind in the trees and to the chuckle of the stream at the side of the road. After ten minutes of walking we stopped to listen to a tinny noise behind us that sounded like someone rattling pots and pans.

I didn't know what to make of the racket so we walked in silence in the shade of the trees where it was cool and where the sunlight only filtered through the canopy overhead. The noise got louder and we stopped again to listen. A small white dog came bounding up the asphalt to stop and sniff at our feet and then it flopped on its side in the shadows of the ferns at the side of the road. In a few minutes, an island boy came walking up rolling a hubcap with a stick and wearing a machete at his waist and a smile on his face.

When he came abreast of us, he stopped and in a very small voice he said, "This is the farm of my grandmother. I can show you," and he picked up his hubcap and disappeared into the jungle that lined the creek.

In a moment we were hopping from stone to stone across the stream and up the bank on the other side. We slipped single-file along a narrow path through the green shadows until we came to a place where there were banana trees, heavy and pendulous, and where the bananas ended there was a breadfruit tree and a huge mango.

"This is the farm of my grandmother," the boy said again. "She come here to pick the breadfruit and the mango. She is old. Sometime she just come to sit with the trees."

We stood quietly and looked around while the boy wandered off to cut free a small stalk of bananas for his grandmother. He tied a short piece of vine to the end to use as a sling and he carried the stalk with the sling over his shoulder. We left his grandmother's farm and walked back to the stream where we sat on the boulders in mid-creek to rinse our feet while the dog bounced from rock to rock chasing shadows. I asked the island boy his name.

"Aschmorre," he replied. "A-S-C-H-M-O-double-R-E, and my dog's name is Blanco," and then he laughed and asked if we were hungry.

He disappeared into the jungle and returned with a green coconut. We sat by the stream while Aschmorre cut the top away with the machete and then we scooped the insides clean and sat on the rocks and ate the soft coconut with the water tumbling through and Blanco splashing in the shallows. When Aschmorre told us he had to go to return to his grandmother, I reached into the daypack for a camera and while he looked away for a moment, I snapped his picture. He grinned at me and then he gathered his stalk of bananas and slipped the loop of vine over his shoulders and hopped up the path to the road. When he left, we could hear the sound of his hubcap rolling down the asphalt, loud like a trash can at first, and then growing dim and then faint and then not at all.

When I see a travel poster in the mall of one of those exotic tropical outposts, I sometimes think of the island of Dominica. I don't dwell on how cool and soothing the shower was on *Sun Flower* or how nice it was to be out of the thrashing trades and to sit in the cockpit at sunrise and drink the French roast fresh from the pot on the stove, or how green the island looked in the soft light brightening over the Caribbean. Sometimes the poster allows me to hear again the trade winds that rustled the branches and leaves of the mango trees and the giggling of the stream that tumbled its way to the sea. Sometimes when it's quiet, I can even hear Aschmorre rolling the hubcap back to his grandmother's house somewhere down that road through the jungle.

On the British Admiralty chart of St. Vincent we found a waterfall just inland from a small cove that was accessible only by boat. When we sailed in, though, the bottom was rocky and foul and there was no safe anchorage. I thought the waterfall might be like the travel posters and so we lodged the anchor between two house-sized boulders about seventy yards out and then rowed a line ashore to tie *Sun Flower* to a palm tree on the beach. We hiked the trail that followed the river inland and you could hear over the sound of the tumbling water a deeper roar that got louder as we walked. The jungle crept in until we were walking through a huge green tunnel with the roar of the waterfall rushing through. Suddenly we were there with the mist and the falling water and the ferns and the icy pool, deep and dark and clear, and all of it in the emerald shadows cast by fronds the size of umbrellas and by the branches above that held a million shimmering leaves.

We stepped into the clearing to watch the water cascading over the edge fifty feet above the pool. I listened to the roar of the water and to the wind rushing through the trees and to the sound of parrots faraway in the hills. It made me realize the years of dreaming and boat building and sacrifice, and even divorce, had filtered all the way down to those few moments on the island of St. Vincent. I kept asking myself whether the journey had been worth it. I leaned against a rock and breathed deeply the air, heavy with mist, and there was a feeling of contentment, like I knew somehow that whatever had gone down before was now irrelevant and that Paradise was right in front of me.

Sun Flower left St. Vincent behind, but I never forgot the waterfall and I never forgot those moments of peace I felt sitting on the rocks in the roar and the spray and in the dark, emerald shadows. We made our way south, though, and after stopping for a few days on Bequia, we sailed on for the tiny island of Mustique that looked on the charts like it might be another dot of Paradise. The sun was low and when we made our approach under sail, we skirted the little shoal outside the bay and tacked over into the shallows where we dropped the anchor in water so clear you could see shadows on the bottom.

I stood and looked about at the low hills on the island and at the white sand beach, but there was a skiff across the way with a trio of local fishermen who were yelling and swearing at me. The men in the open boat had just paid out their nets and when I tacked to port entering the bay, we sailed over part of their set. I kept looking over toward the fishermen. They yelled and shook their fists and started to haul in their nets and so I went below and grabbed a bottle of Mt. Gay and dived into the water to swim over to them. When I pulled myself aboard the fishing skiff, I handed them the bottle of rum and apologized. They took the bottle from me and only sat and scowled. There was no gratitude and they didn't want to talk so I told them to keep the rum and that I hoped the fish would return and then I dived into the water and swam slowly back to my boat. It seemed like something bad had happened to Paradise.

Later in the evening we heard music coming from an open bar in the village. We rowed ashore to listen to the scratch band play the island reggae and to dance and to drink bottles of Heineken fresh from a barrel of ice. The fishermen were there and when I entered the bar, they glanced my way and frowned and muttered to each other. I remember thinking what jerks they were to let a simple mistake bother them so. The reggae took over after that and everyone danced and drank beer. The lady I was with had a terrific time with all the locals who had fallen in love with her. One by one the guys from the fishing skiff came over and they talked to me about their lives on Mustique and I told them of my life of sailing and how I had built my own boat. When we left the bar in the early morning, we parted as friends. I told them that when I came back to Paradise, I wanted to go out with them to help set the nets. They invited me along, but only if I would bring another bottle of Mt. Gay.

Travel posters might capture the color and the form and the shadow and light of places like this, but never the mystery or the essence, and maybe that's why so many of us can sit and stare at them. The posters only hint of the real thing. You sit and look and let your imagination try to fill in the soul of the place, and all of the doctors and lawyers and businessmen who studied while you were in school will walk by and see you sitting there. They will hurry on their way to work to hustle their money and still you sit there dreaming and looking out through the windows, wondering if any of them would ever understand what you had come to know, that the travel posters and the dots on the charts and the roads marked by broken lines, and all of those coffeehouse windows, are really just doors that open into another kind of Paradise.

/) CHAPTER THIRTY /)

Herbert...

San Diego doesn't get a lot of wind so you can hear the parrots in the Point Loma neighborhoods screeching down the canyons from a long way off. The sounds don't belong here somehow, like the parrots are only passing through from a land of heat and mangoes hidden somewhere over the horizon. At first I thought a breeder set them free each day to exercise, but parrots have no homing instinct. It's apparent they are not only thriving here but reproducing as well. Now the flock is so big and noisy, people turn their heads just to see. The parrots fly in flashes of green along the streets and over the stucco jungle and when you hear the echoes, they make you think of the tropics of Adventures past where birds weren't trapped in asphalt cages.

When the cold Northers stopped blowing and the trade winds drifted south, Key West settled into small town softball games, Fourth of July picnics, and a tropical quiescence disturbed only by the nightly flyovers of the mosquito plane. Summer also brought with it the faint aura of Paradise, as if the Key West of Hemingway hid just down the street in the shadows of the magnolias that spilled over the sidewalks. There were no mangoes and no papayas and the only wild parrot was the bar on Whitehead Street, but a kind of timelessness drifted through the town asleep in the summer heat.

The back street neighborhoods were always quiet, yet it seemed like flocks of parrots should have been hiding in the high palms that waved those lacy fronds overhead. The only echoes you heard were from the cars and the honking horns up on Duval Street.

There was one parrot, though, that squawked from the open window of a tiny conch house over on Caroline Street. The bird's owner used to shout for Herbert the dog each morning to come to the porch for a bowl of kibble. It didn't take very long for the parrot to start mimicking the owner. When I stopped for a cup of Cuban coffee at the kiosk at the end of the street, you could hear the parrot yelling "Herbert" all the way down to the corner. One morning the dog was killed in the street by a car that didn't stop. The parrot never knew the dog would not come back and for months afterward, the early morning cries of "Herbert... Herbert" echoed down the streets and through the shadows of Hemingway's town. The parrot was still yelling for Herbert when I left to sail into the sunset the following spring.

Down on the island of Nargana in the San Blas Archipelago, a little man in a coffee kiosk told me there were parrots on the Rio Diablo that flowed from a spring deep in the jungle of the Darien Gap just across the channel. After a grueling three-hour row in the dinghy, I reached the spring without seeing a single parrot. I sat motionless in the little boat and drifted in the shadeless river, shrinking into myself like so much copra in the furnace of midday while the trade winds lay smothered beneath the sprawling, listless sun. Sapped of my energy, I sagged into a near catatonic state and resisted any kind of movement, conscious only of the heat and the quiet and the river that wandered through the jungle to carry me back to the sea.

At some point I heard the sound of parrots in the distance. I didn't see them at first but suddenly a patch of green burst from the top of a mango tree that slid by in the stillness. More parrots soon followed, flashing their colors in the sunlight, unaware of the outsider in the dinghy at midstream. They continued their search for the perfect mangoes, leapfrogging from tree to tree and chattering in a noisy raucousness that echoed beyond the bends of the river. There were dozens of the *loros pequeños* flitting about in a lime green cloud that swept in and out and over the branches of the trees, but then they left as suddenly as they had come. When the jungle silence returned, I receded into the heat and the quiet and drifted back to the sea with the sound of the parrots echoing in my head.

Down the streets of Point Loma the chattering has come back, brought to life by a flock of misplaced parrots searching for mangoes that don't exist. They have as good a chance at survival as the parrots that live in the jungles of Chicago, but I worry just the same. All I have left to remind me of the tropics is a bunch of withering house plants and an album full of aging pictures. When I sit with my morning coffee and listen, I can hear the parrots up the canyon not far away. For a moment it seems like I could be sitting in that dinghy again on the river in the Darien Gap. If the parrots disappear, the pictures won't fade and the plants won't die and the coffee will still steam in my cup, but the early morning squawks will be pierced with silence and I'll be left alone again, stranded like the parrot in Key West calling for Herbert, the dog that never came back.

CHAPTER THIRTY-ONE

staring from the rafters...

There were no serpents or uncharted reefs when we left Grenada, and on the morning of our sixth day out we could see the hills of the Darien Gap in Panama and then the tiny, palm-lined islands of the San Blas Archipelago. We headed toward the pass in the reef and slipped through the narrow opening to enter the calm water inside where the white sand sparkled in the shallow sea. I looked around at the islands that were low and smothered in coconut palms, scattered about like green tufts on a sheet of blue silk. I hadn't thought of Janie very often, but the tiny islands floated by and I remembered what it was like with her so in love with the cays of San Blas when we had been there some years before.

Isla Tigre lay among the islets to the east. We jibed the boat over to clear a reef that extended beyond an unnamed cay and then turned up to cross the thin water of the banks. Buddy and Marianne stood near the bow rail watching for coral heads, but the sandy bottom extended for miles. They stared at the transparent water and at the shadow of the boat and at the little islets slipping past.

Sun Flower rounded up in the reach just to the lee of Isla Tigre. We hadn't been stopped more than five minutes when a dugout approached from the island and coasted to a stop on the starboard side. The *ulu* was filled with people that turned out to be an entire family coming out to greet us. Several dozen *molas* were soon draped across the dugout and for ten feet along the lifelines. There was another ulu approaching so we sent the first family on its way. I waved the next family off and then Buddy and I launched the sabot before any others showed up. We made our way ashore and stood in the main street of Isla Tigre looking at the immaculate village of thatched palms and sandy streets swept clean and at the tiny Kuna Indians smiling and holding out more molas.

We were invited by one of the village elders to join him for a snack of plantains roasted in an open fire pit in the center of his home. The interior of the thatched house was bright and fresh and there were some furnishings made of coconut logs scattered around, but there wasn't much inside except for all the dolls hanging high along the walls. In broken English the Kuna family told me they liked to trade their molas for children's toys, especially dolls, and there were dozens staring back from the rafters. I kept looking at them and thinking of Janie who loved the Kunas and loved all the dolls, but these were old and faded and there was a layer of grime on their arms and shoulders and all the color was gone, even from the faces.

We ate the plantains and bought three more molas from the Kuna family and then left and walked the length of the main street. We ducked into several more houses where the families shared their molas and where the dolls stared at us from the rafters. I didn't like the blank faces and the hollow eyes. When we were invited into yet another home, I stayed out in the street. Buddy emerged from the shadows of the entryway and walked over to where I stood on the corner.

"I don't want to spoil anything, Buddy," I said. "It's tough sometimes."

"We don't have to stay."

"I didn't think it would bother me."

"Let's leave and sail out to the other islands where we can get in the water."

We turned to walk back toward the rickety dock where the sabot was tied and then rowed out to *Sun Flower*. The exhaustion from the trip set in and the four of us were asleep as soon as we stepped below into the quiet of the cabin.

I woke right at sunset and pulled the curtain away from the portlight just to see the colors. *Sun Flower* wasn't swinging at anchor much because of the current through the reach, but my window looked out to the west and I saw when the sun outlined the mountains of Panama and turned to pink the clouds stacking against the hills. Twilight settled in and I rolled from the wingdeck berth and stepped out into the night where the lights of Isla Tigre shimmered on the flat water. On the west end of the island there were brighter lights and it seemed like a festival might have started, but I sat on the cabin roof and leaned against the mast where I shut my eyes and thought again about Janie.

In the morning I started the diesel while Buddy drank his coffee down and then went forward to haul in the anchor. We drifted down the reach in the current and then motored around a sandy cay and onto the open banks. Buddy and Marianne and Chrissie got the main and jib up and drawing and we scooted away at ten knots again to cross the shallows toward the surf booming in the distance.

We found a spot to anchor near a horseshoe islet where we spent the day diving on a fairyland of coral. We moved on to an island close to the barrier reef where the diving was good for grouper and lobster. After an encounter with a half-dozen six foot sharks that couldn't be driven away, I found the spot on the chart where we could sneak through a pass in the reef and get out into the trades that were rushing through, fresh from the east.

The following day we made our way into the narrow opening where *Sun Flower* fell away and began surfing up to fifteen knots and then falling back to ten, then surfing to fifteen again when the big swells rolled through. We passed Punta Manzanillo in the afternoon and then made it around into Portobello. The sun sank low in the sky behind the mountains, but there was enough light for us to stand on the cabin roof and look at the ruins of the Spanish gun emplacements guarding the entrance to the bay and at the fort itself that formed the foundation for many of the buildings in the town.

We rounded up right at dusk and dropped the anchor just off the old ramparts where we could see from the deck all the old cannons strewn about and the walls that were still intact. The Panamanian government had spent money to clear the jungle from the grounds around the fort and on the points of land at the entrance to the bay. The whole place looked like a national park except that *Sun Flower* was the only boat in the anchorage.

We were in the wind shadow of the hills to the east and as night settled in, we heard the screams of the cicadas in the trees beyond the fort. I thought the anchorage so close to town would be buggy and full of mosquitoes, but the insects never showed up until after we had launched the sabot and rowed ashore for a quick tour. After looking at the cannons and the turrets and the grounds of the fort, we had enough of the bugs. We rowed back to *Sun Flower* where the rum and beer waited.

We spent a quiet night ghosting on the anchor line while the soft breezes that swirled from the hills above were just strong enough to keep the bugs away. When the sun climbed high enough to heat the cabin, I stepped outside to see Portobello again in the dim light of the morning. There were people walking by and a pair of wiry dogs sniffing along the shallows of the bay and a lone flatbed truck idling along the road from Colon bringing a load of building supplies. I watched the people, but they never once looked out into the anchorage. I felt like I was snooping. I went below to help Chrissie start some coffee. It didn't take Buddy and Marianne very long to make their way out of the forepeak.

We finished a breakfast of fried Danish ham and some diced potatoes that were nearly bad, but the coffee made up for it. I stowed the dishes and then we piled into the sabot to go ashore and look around the little town. There wasn't a lot to Portobello, just a store or two and a few small houses and a small white church and some outbuildings, some of them perched on the old fort walls. We wandered into the little church where Marianne and Chrissie kept staring up at the statue on the cross. Buddy stood there scratching his beard.

"Son of a gun," he said.

"I love it," Marianne said. "Can you imagine what people back home would say?"

I looked for a long time at the figure, jet black and shining in the light from the open doors.

"If there is anything like this on Judgment Day," Buddy said, "I wouldn't want to be from Mississippi."

Chrissie touched my arm and then she looked back at the small ebony figure hanging on the cross.

"It's beautiful," she said. "Most people want to see only what they believe, though. Buddy's right."

"I think so," I said.

Buddy and I left the church and walked over to where the guns were mounted on the hills to guard the entrance to the bay. We could see across the water where the Panamanians had cleared away the jungle growth to expose the ruins and some towering mahogany trees near the top of the hill. We hiked back along the road from Colon and went over to the main part of the fort where we climbed the ramparts and looked at the cannons laying against the walls and at the interior where the grass grew thick and green along the corridors. There were lookout turrets and walls blackened by moss and stairways where the steps had been worn down by the boots of the Spanish soldiers. Buddy hesitated and looked back at the town scattered among the dark gray walls and then he looked over at me. He caught me staring past the walls and the ramparts and the turrets to the open Caribbean beyond the mouth of the harbor.

"I'm sorry about Janie."

"I know," I said. "Me, too."

CHAPTER THIRTY-TWO

the fawn...

It's hot and windy this afternoon. There are seeds that look like puffs of down floating over the parking lot, rolling like tiny tumbleweeds across the asphalt. They make you think of soap bubbles at a preschool and pillow fights and the band music of Lawrence Welk. There are a great many of them outside the windows of the coffeehouse, so thick they form miniature drifts against the curbing and then swirl again in the wind when someone walks beneath the eaves. If you watch long enough you can see them high in the air like helium balloons that have come loose, but down along the sidewalk they drift and swirl and scatter. It's like someone just took a deep breath and blew on a thousand dandelions. One of them floated in front of the window in the ebb and flow of the air currents. I watched it absently and thought about motes of dust and Phyllis Diller's hairdo and then it swirled in the wind and was gone so fast I wondered if it had been there at all.

I stood in a field this afternoon and heard the cry of a hawk. When I looked to the sky, I saw a pair of them, one of which was carrying a small animal in its talons. The hawks were adult red-tails. They flew in lazy circles opposite each other and looked so much like sailplanes I got the idea they were showing off. I watched them for a long time to see where they might land, but they disappeared over the hill to the east. I looked toward the crest to see if they might return. Somewhere beyond the hill there may have been a nest and a young hawk waiting. I didn't see them again.

At one point during those Myrtle Beach days, I stood in the kitchen making a jelly sandwich when a hawk flew into my backyard and landed on the branch of a tree only a few feet away. It gripped a dead mockingbird in its talons and stood tall on the branch of the pine and looked about. It was so big I thought it could have been an eagle. My heart started pounding and I got so excited I couldn't breathe. I bent down and crawled away from the open window to get my BB-gun from the closet. I sneaked back into the kitchen hoping to shoot the hawk out of the tree, but I watched as it buried its face into the body of the mockingbird. When the hawk lifted its head, you could see the bloody entrails hanging from its beak. I stood there with my BB-gun cocked and ready, but I could only hide and watch as the hawk fed again and again on the mockingbird.

After a minute, some movement caught the eye of the hawk and it spread its great wings to glide away from the limb and away from my backyard and out of sight beyond the pines that towered over the street. I stood at the kitchen window and took a deep breath and stared at the empty branch of the tree. I put my BB-gun on the counter and reached for the jar of jam. I felt an odd sensation then. It was like I couldn't make up my mind if there really had been a hawk or if it had just been a moment of fantasy.

On my first trip to Panama, I got to know one of the Kuna Indians who lived in the San Blas islands on the north coast. I asked him one day if he knew where the big grouper were hiding. He told me to join him early the next morning and to bring the biggest spear gun I had. At dawn we loaded our gear aboard his *ulu* and headed for the reef where you could hear the booming of the surf a half-mile away. Denny kept telling me the giant grouper were just beyond the line of combers that crashed over the coral. I wasn't sure we would make it through without capsizing, but we sailed parallel to the swells and darted into the open sea when there was a lull. One hundred yards beyond the breakers, Denny headed his dugout into the wind and let the heavily patched main luff and slap and shake. I furled the sail while he dove into the clear water to tie the boat off to a coral head ten feet below. When he surfaced, he told me to swim toward the reef line. I would find a grouper, but I needed to be careful because the fish were so big they could swallow a man. I smiled and then flopped into the water with the spear gun and headed for the wall of coral in the distance.

Beyond the pass through the reef I could swim in the broken surf without being pounded, but a big comber rolled through and caught me by surprise. The foam and the bubbles cut the visibility to zero. I floated with my head down and waited for the water to clear. When I could see again, a brute of a shark not ten feet away swam in slow motion toward the pass. I could see the ice in its eyes. I waited again to see what it would do when another comber rolled over us and I could see nothing but foam and bubbles again. I thought the shark might explode through the foam and tear at my legs. My heart pounded and I couldn't breathe. I bit the mouthpiece of the snorkel so hard I thought I might cut it with my teeth, but the foam and the bubbles cleared once more and the shark had disappeared. I restarted my breathing and swam back toward the ulu wondering where the shark had gone. I remembered what Denny had said about the grouper.

To my right along the wall of the cut a series of coral heads looked like lumps of granite. I swam toward them looking for a grouper while I kept an eye out for the shark and tried not to shudder in the chill. I didn't see the fish until it turned broadside to me. I realized then it was not a coral head but a 400-pound grouper, slow and gray and huge. I aimed my spear gun at the spot where its brain would be. I followed its movements, but I was shuddering inside and

I didn't want to pull the trigger. I watched as it flicked its massive tail and swam away into the gloom.

I told Denny about the grouper while we sailed back in the ulu. He laughed and told me it was a good thing I didn't shoot it. I would've been towed all the way to Jamaica if I had missed its head. If I had only nicked it in the side, I would have drowned trying to get out of its mouth. The thought of these possibilities made Denny laugh even harder, but then I told him about the shark. He only looked at me. I went to a slide presentation once where the narrator had slipped into the middle of his program a picture of a beautiful girl in a bikini. We watched slides of sunsets and slides of mountains and slides of rivers and then suddenly there was this girl and everyone in the place woke up, but she was only there an instant before another slide of the Grand Canyon came into view. When it was over, I wasn't sure if there had been a girl at all. I didn't think Denny understood, but he turned and stared toward the reef in the distance and then he nodded his head.

"Always there are sharks," he said. "You are lucky."

I was once placed under house arrest in Honduras and spent one of the afternoons sitting in a tiny infirmary on the Navy base in Puerto Cortes. My right foot was infected and swollen and I couldn't walk. I sat on a cane-back chair near the medicine cabinet in the corner and waited for the doctor. I was alone in the office. An open door yawned near the opposite end of the room and I sat listening to the sound of the trade winds sifting through the branches of the trees beyond and to the flies that were buzzing against the screens that covered the windows. There was some slight movement in the room. While I sat motionless in my chair, a fawn stepped through the door on feet so light you only thought of down. Minutes ticked by and then it began to place its feet just so to cross the room. The fawn inched toward me and leaned far forward like a curious and tentative puppy leading with its nose, cold and wet and quivering, to touch my hand on the chair. If I could reach right now through the window of the coffeehouse and catch one of the puffs drifting by, it wouldn't be any softer than the touch of that fawn. I didn't move, but the fawn backed away slowly and then it watched me for a moment longer. I could see in its eyes the innocence and the wonder, like looking through tiny windows into another soft, dark, and infinite universe. I wasn't conscious of time just then, only that the fawn backed from me once more and turned and stepped like so many puffs of down through the door and into the jungle beyond. There was nothing left in the infirmary but the sound of the trade winds again and the noise of the desperate flies pinned against the screens in the windows.

A gust of wind has blown fresh against the windows. There are hundreds of the seeds swirling beneath the eaves. Puffs of down are everywhere and it makes you wonder how many will ever find a spot where there is enough soil and light and water and how many of them will not be the lucky ones and just disappear. They are beautiful in flight, like looking through one of those little glass globes at Christmas that you shake to get the flakes of snow to fall quietly over Santa Claus. There will be another gust here soon to take the puffs away. It will seem like they never were.

After my second trip to Panama, I hit a reef and tore the bottom out of *Sun Flower*. I had to leave the bones of my boat scattered over the coral heads in one of the most remote areas of the Caribbean down off Honduras. There was a lot of pain in that country. The wind has come again and the puffs of down outside have disappeared. I have often wondered if there ever was a fawn.

CHAPTER THIRTY-THREE

Sara Jane...

I glanced toward the parking lot when a whale-tailed Porsche Turbo roared in and chirped to a stop in front of the coffeehouse. A smallish man in his thirties sporting two-inch lifters in his heels stepped out of the car. He let his door swing so wide it hit the Toyota in the adjoining slot. He never once looked at the new prang in the other car. He shut the door and went hustling off with a smirk on his face like he was late for his own meeting. You have to wonder about a man who drives a $100,000 automobile like it was a banged-up '72 Pinto. I suppose if you are rich enough it doesn't matter what you do with your toys.

I worked for a while in South Florida as a boat builder. I came to know a few dope runners, mostly single guys in their twenties who had so much money they only played at life like the guy out front and his expendable Porsche. A lot of them had offloaded bales of high-grade Colombian from some worn out coastal freighters onto private yachts down in the low islands of the Bahamas. They ran these boats right up the New River in Fort Lauderdale. There were so few federal agents around, the dope runners could bring the loaded boats up the river in broad daylight on a Sunday afternoon and not be stopped. The skipper of one of those yachts could make enough money in one day to pay cash for the Porsche parked in front of the coffeehouse. When enough door dings accumulated, he could make enough on subsequent trips to trade the Porsche in on a Ferrari. Some of the guys I knew did just that.

Thirty years ago things were a little simpler than today. Most of the people in the marijuana business in South Florida didn't even own pocket knives. They spent a lot of time in the chrome-and-mirror bars on the Fort Lauderdale waterfront, laughing and squeezing the girls hanging around who always were squealing and laughing and slobbering over the dope runners. None of them ever worried much about being caught except a good friend of mine who drank himself silly until the stress and the alcohol made Swiss cheese of his stomach. He had to go in to get the holes stapled shut. After his surgery, Joe didn't drink for months, but then he couldn't stand the anxiety again. He wandered off to a neighboring town where no one knew him and binged for three days. When Joe came back, there wasn't any color to his face and his lips had a blue tint. I didn't think he could live through much more.

His business partner had just the opposite personality. With Len there was never a thought of being caught. Every day was a carnival for him and he was the clown with the calliope. Laughter followed him around in peals of happy thunder. He told me a story once of how he and Joe had loaded a U-Haul trailer full of Colombian and then towed the trailer across the country. Somewhere in Indiana their car broke down and the whole rig had to be left at a garage where the service bay wasn't long enough for the car and the trailer. The mechanics simply lowered the garage door until it came to rest on the rear hitch and then locked it in that position with the car in the bay and the trailer full of marijuana left outside in the open. Len slept like a baby in the motel room. Joe drank himself silly again and worried and fretted all night, terrified the police would come crashing through the door of their room at any moment. He never went to sleep.

Len and Joe had been sailing together long before the dope running business began. After they made their first fortunes, they continued sailing and soon owned a small fleet of yachts including a famous Swan 65 that had been raced around the world. The boat had been sponsored by a huge watch-making firm during its racing days. Len constantly joked about it, wondering what the Swiss company would have done had they known their pride and joy was now filled with marijuana bales half the time. To the guys' credit, though, they enjoyed sailing and so they used the boats a lot, whether it was just for day sailing or for sneaking off across the Gulf Stream to Nassau to retrieve some of the gobs of cash they kept in secret numbered accounts.

I had the opportunity to work on some of the boats in the fleet. My favorite was an old wooden trawler that had been built in Ireland. *Sara Jane* had massive scantlings with six-by-eight timbers for her deck beams and a hull so heavy and solid there were no sounds to the sea below decks, only the dull throb of her diesel. She was a boat built to take the cold North Atlantic with precious little about her that was soft or plush. She was never a yacht so the spoiled dope runners wouldn't stay aboard her for long. She could take the sea and bring you home, though, and there was no pounding or rolling or shuddering, just that feeling of power and strength you get with good and solid workboats.

Sara Jane was used once as a dormitory in Nassau when Joe decided he should marry his long-time lady friend. All of the dope-running fraternity were invited to the big wedding in the Bahamas. I was part of the delivery crew that took the old trawler into Nassau for the big day. After we anchored fore and aft, the other smugglers who had crossed over the Stream rafted up alongside until there was a continuous line of yachts. The dope runners could hop from boat to boat and eat and drink and party in Joe's honor. Everyone had a grand time of it with the champagne and the *hors d'oeuvres* and the music. There was so much giggling and laughter going on that your face got tired of holding the smiles. You had to find a quiet place on *Sara Jane* just to gather your composure.

I stood on the afterdeck during one of those quiet times and looked out over all the dope smugglers hopping here and there on the million-dollar string of yachts. All of them were laughing and singing and laughing again, and on into the night went the laughter. Some of the laughter came with tears and then there was bellyache laughter and then nighttime-girl squeals of laughter and then gagging-and-retching-and-rolling-on-the-deck laughter, and on and on went the laughter of the dope runners. I wondered then, from the quiet afterdeck of the old and solid *Sara Jane,* if any of them ever heard the echoes.

CHAPTER THIRTY-FOUR

another shot of Screech...

Newfoundland is cold, even in August. I ducked into a street café just beyond the quay in the harbor of St. John's to warm up with some coffee because the wind off the water cut like a rapier through my windbreaker. There was an old man inside sitting at the counter reading a newspaper. After I got my coffee, I watched him eyeing the waitress. She was chewing with big bites on a piece of toast and leaning over the counter to point things out in the back pages, but the old man kept trying to see down her blouse. The waitress was built just like Bette Midler and I was envious. I watched them from a distance and sipped the black coffee.

I thought about moving to the counter, but through the window toward the harbor I could see the masts of a fishing schooner hidden just beyond a small freighter tied to the quay. The topmasts had been cut away but the spars of the schooner still towered over the waterfront. I decided to finish my coffee and walk down to see. The waitress who looked like Bette Midler came over to ask if I wanted a refill. She didn't lean over my table. I left her some money and zipped my jacket together and smiled at her as I walked through the door of the café into the cold and the wind.

The schooner was old and rough and not very pretty. I stood near the high bow to look down her sheer, but the hull was so hogged you couldn't picture the sweeping curves that used to be the lines of her hull. The original deckhouse had been torn away and replaced with a box, square and sharp and much too big for her lines. The schooner appeared too tired to work much longer. I didn't stay long in the wind to look at her but walked, instead, down the harbor toward the gangway that angled from the side of the little freighter made fast along the sea wall.

On the mast over the bridge of the ship, a hammer and sickle on a red background stretched itself to leeward and quivered in the wind. I stopped to look at the flag and at the freighter when four Soviet crewmen walked down the gangway and turned in the direction of the café. They were only a few feet in front of me. I felt awkward being so near, but they didn't look like Cold War enemies of democracy. The four of them turned to stare at me and for an instant I thought one of them might stop, but they kept walking faster toward the café. None of them bothered to speak. The four sailors disappeared beyond the quay on their way to town and I was alone again in the wind.

I didn't feel much like going back to an empty hotel room so I walked the length of the quay and climbed the hill on the north side of the harbor. You got a good view of the narrow ship channel on the way up. A small cutter, maybe a thirty-footer, tacked into the wind blowing icy cold from the open sea. The sails were white and pretty and clean and I watched the little boat turn about and tack toward the north side, but there were moments of sadness sweeping through, like the gusts of wind that snapped at the trailing edge of the cutter's jib. *Sun Flower* would have been footing at twelve knots or better down there, but then there was the reef in Honduras where she was torn to shreds. I turned my back to the cutter in the channel and looked off toward the freighter and the masts of the schooner in the distance. The reef in Honduras seemed as cold and sinister as the Newfoundland wind.

I started to walk down the hill toward the town when I passed a small frame house painted in Caribbean blue. I stopped for a moment just to look. A woman was sitting on the porch peeling potatoes. I hadn't seen her on the way up. She smiled at me and waved.

"You a Yank?" she asked.

"Yes," I said. I wondered how she could tell.

"Hope you like St. John's. No place like it."

"It's beautiful. Cold, though."

"Breezin' up today, but you get used to it."

"I think I spent too much time in the tropics."

"Go on down to the first pub you see and ask for some Screech. It's some jeesly good at keepin' you warm."

"I'll try that."

"Try some cod tongues, too. You'll be an official Newfy then."

I waved to the woman who stood and smiled and then she stepped back inside the spotless blue house. The gusts of wind blowing in from the North Atlantic still made me shiver and I thought a shot of Screech would be a good idea. I turned again and walked down the hill toward the center of town where there were smiling people on the sidewalks and cars parked along the cobblestone streets. I sneaked into a darkened pub and took a seat near the far corner. In a few minutes a barmaid came over and I ordered a shot of Screech and a bottle of Molson for a chaser. The rum felt like fire going down, but the woman in the blue house was right. It didn't take very long to forget how cold it had been in the wind.

The barmaid came back and asked if I'd like another Screech. In a moment of weakness, I nodded and smiled. The pub made me feel like I was back in Port Royal drinking the good Appleton rum with *Sun Flower* tied to the dock and the hot wind sweeping down the mountains and out the harbor entrance. The second shot went down in another cascade of fire and made the Molson taste like old tonic. The barmaid giggled and took the empty glasses away. I sat for a moment while the warmth surged over me like bathwater. I folded some money and slipped it beneath the empty beer bottle on the table and then left the pub to walk back along the quay.

The Russian sailors were gone and the freighter sat tired and still, tied fast to a pair of thick, steel bollards fore and aft. I watched as a steady thread of smoke fell away from a small stack near the aft end of the bridge. High overhead you could see the hammer and sickle stretched tight, still shaking nervously in the wind. I kept looking at the flag and then at the empty gangway and I wondered if the four sailors were aboard. I waited for a few minutes thinking someone might show up, but in the end I walked well past the worn out schooner before I turned to look again at the freighter.

The Screech made me numb to the cold for a while, but the wind began to cut through again and I hunched my shoulders and walked past the café toward the hotel that brooded over the center of town. My room was on the fifth floor and when I unlocked the door, I walked over to the balcony and looked down on the street far below. People like ants were moving about, clutching their throats in the wind, and there were flags snapping away over the building across the street. I stepped back into the hotel room and shut the door behind me and then flopped on the bed and held my hands over my face. The rum made my eyes swim. *Sun Flower* was there again, fading in and out, while Janie stood alone on the dock laughing. Newfoundland is cold, even in August. I wished I had ordered another shot of Screech.

CHAPTER THIRTY-FIVE

feathers of spruce...

We walked up the street from the destroyer pens in Gibralter to find a café where we could try the local coffee. The narrow streets were lined with shops and we didn't walk very far before we found a café owned by a young Spaniard with an English accent who told us he served only the finest coffees and teas. We ordered two mugs and took a seat at a small table near the window. The coffee showed up in those thick, porcelain cups you only find in the Navy, but the cups were so stark and white they made the coffee look like roofing tar. The coffee didn't taste like tar. We each had a third cup while the kid grinned from behind his counter.

It was cool in the shadows of the café, but we needed to get back to the boat where Pudge and Marcia waited. We walked back toward the pens to the slot where *Cimarron* was tied. John tried to jump to the boat from the high walls of the dock by grabbing the mast and sliding down toward the deck, but his foot slipped from the halyard winch and landed square on the cleat below. The gash went deep into the ball of his foot and in a moment we were back on top of the destroyer pens limping toward the shops and the narrow streets again and the medical clinic a block past the morning café.

A nurse in the doctor's office gently cleaned the wound and smiled at John while she worked. Instead of worrying about the gash in his foot, John fell mildly in love. The nurse had a way of watching him that made her look like she was smiling from behind some Spanish lace. She had eyes the color of Amantillado and a complexion so smooth you thought of the soft white linens resting on the shelf, but the doctor was too fast and efficient. John could only glance at the nurse as the doctor ushered us out the door. While we hobbled back to the boat, John stared down at the cobblestones.

By the time we reached Malaga, the cut in John's foot had healed enough for him to limp without crutches. We left *Cimarron* behind to travel through the rest of Spain to search for a classical guitar. I didn't speak Spanish and John knew just enough to be misunderstood and so we trudged the streets of Malaga and roamed through the music stores looking for someone who could tell us where to find the artisans who made instruments for the world's great guitarists.

The directions we were given always sounded like gibberish. We wandered over miles of cobblestone streets only to find another retail shop at the end of our trek. We had a list of names that had been given to us by one of the Romero brothers, but he hadn't written any addresses. Each time we asked a salesman where to find these distinguished people on the list, the salesman pretended to know exactly and then sent us limping on our way across town. We wound up at the doors of another guitar shop where the new salesman laughed at his compatriot's stupidity and then sent us limping in another direction.

Summertime heat gets trapped in the canyons of the back streets of Spain. When the ball of your foot is mangled and the cobblestones bake in the afternoon sun, it isn't easy to walk. We couldn't afford to hire a taxi so John moved in a kind of osmosis and we trudged around and saw a good deal of Spain in slow motion. We stopped and rested a lot to try to keep the swelling down, but there was so much futile walking in the heat that we were ready to quit. Once when we found one of the guitar makers, the man was old and slow and arrogant and he wanted a $2,000 deposit on an instrument that wouldn't be made for another five years. Neither of us thought the man had five years left to live. We looked at his unremarkable guitars and then trudged out the door.

There was a café down the street from our pensión. After our arrival in Grenada, we stepped into the shadows inside to have a beer and to figure out what to do next. There were bullfight posters taped to the walls and casks of sherry stacked on the floor. Behind the bar hung a black and white picture, autographed in red ball-point pen, of a very tired Ernest Hemingway on his last visit to Spain. The beer was good and I stood for a long time in the low light thinking about Hemingway while John roamed along the walls and tried to read the bullfight posters.

The owner of the café, who was tiny and old and very tired himself, came over to me and asked in broken English if we might want to try some sherry. I didn't like sherry, only the kind my father bought at Safeway for ninety-nine cents a bottle that he drank out of a yellow tin cup filled with ice. The tired little man pointed to a wooden cask behind the bar that was stamped "1925" and something else in Spanish above the word "Jerez". John came over when the bar man poured from the cask two glasses of sherry. After our first sips, we understood then one of the great gifts to the world from Spain. We left the little man behind the bar and limped our way back to the pensión, but our time in that country had changed. We hunted for guitars and stopped often to sample the sherry and it made us feel a part of Spain then, like we were no longer looking at the country through tourist eyes.

The little barman had suggested to us we might find the men who made the guitars if we talked not to the people who sold them but to the people who played them. He told us of a man who could be found in the square in the evenings who strummed Flamenco for tips and for glasses of sherry.

The next evening we limped over to the square where we heard the Flamenco guitar as it echoed down the streets of cobblestone. We sat and listened and put coins in the man's case and waited for him to take a break from his work. John asked in halting Spanish what kind of guitar the old man was playing. The musician wrapped his thin fingers around his chin and looked at us for a moment and then he showed us his guitar. The label inside the sound hole was signed by Antonio Marin, one of the instrument makers on our list. When we asked the Flamenco player where we might find the man who had made the guitar, he wrote an address on a cocktail napkin. John folded the napkin into his wallet.

By mid-morning the following day, the sun was full in the sky. The heat had already settled in and we had limped a long way on cobblestones, but we finally found the street written on the cocktail napkin. We were both tired and dusty and thirsty. I was a lot more interested in finding a place in the shade to sit and drink sherry. We walked up and down the length of the narrow street again and again trying to find the guitars of Antonio Marin.

The sky looked hazy and bleached in the heat and the buildings on the street were the color of the hills that baked beyond the city. None of this looked anything like where you might find one of the world's great instrument makers, but on the fourth trip up the cobblestones I passed close to a low building that faced south toward the sun. The window shades had been drawn to ward off the heat. One of the shades wasn't fully closed and I looked beneath into the gloom where I could see on the far wall the patterns and the forms they use to make the backs and sides of guitars. I turned and shouted up the street to John that it was here after all.

I tapped lightly on the window. A man looked out and then motioned to me that he would unlock the gate to the courtyard. When we were invited inside, John took from a stand one of the guitars that had just been finished. It was so perfect an instrument that he didn't want to play it. Antonio Marin insisted and I sat and watched and listened as John played his scales. You could see in his face that he had only played guitars like this in his dreams.

I took a seat in a corner of the shop and watched while Antonio Marin drew a flat blade over the spruce top of an unfinished guitar. The blade went true and straight along the grain and the spruce peeled away in sheets so thin they were translucent. He took the instrument in his hands and looked across the top and you could see it was so smooth and fine the light from the lamps was reflected off its surface. He eyed his work and then set the guitar down again and then slowly, with strokes long and straight and true, he drew his blade again along the grain of the spruce. Delicate shavings floated to the floor like feathers.

John would have paid for a guitar on the spot, but as we were to discover at many other places in our travels in Spain, the artists who make these guitars have customers waiting all over the world. It would be years before the orders could be filled. When we left the shop in the afternoon, a very kind Antonio Marin gave us the addresses of the other guitar makers. We limped our way through Spain and looked at the instruments meant for the Parkenings and the Romeros and the Breams. The experience made us realize we were only on the outside once again, staring through the windows. We would forever be the ones sitting in the audience and the guitars we saw were only meant for the people who played so well you became one with the soul of the guitar and with the soul of the musician looking down on us from the center of the stage.

We spent a lot of time in the heat and in the dust of the back alleys in Spain looking for those guitars. In later years when we attended concerts together, the music we heard carried with it the souls of the artists who worked in the little shops on the cobblestone streets, the ones who drew the blades straight and true while the curls of spruce floated in the air like the notes that floated from their perfect guitars.

We also knew, after finding Antonio Marin, the great guitar hunt in the summer of '78 was not the hunt we had planned. John finally wandered into a shop in Madrid where they sold guitars for Jose Luis Ramirez and for Manuel Contreras. He bought a lovely Ramirez classical guitar that he has enjoyed for the last thirty years. The notes still drift and float when he plays. I can sit in his living room and listen with awe to the music of the masters, but when I hear the notes from John's guitar, I often think only of Antonio Marin drawing his blades along the grain and of the notes he played that drifted softly in the air, the lacy feathers of spruce that floated like whispers to the floor.

/) CHAPTER THIRTY-SIX /)

real palmiers...

We caught a good train in Madrid but when we got to the frontier in Hendaye, we had to transfer to a French cattle car for the all-night run into Paris. We passed a great many freight trains heading the opposite way and the noise and concussions from two locomotives passing in the night shook and rattled the old French cars. We couldn't sleep until we neared the outskirts of the city. We were in desperate need of coffee by then. The train hissed to a stop in Gare d'Lyon so we left the cattle car behind and walked a long way up the street where we sat outside at a café and ordered one coffee and one *palmier* because we didn't have enough money for two.

The palmier was crisp and sweet and I munched my half slowly and washed it down with a swallow of raw French roast that scalded my throat. There were people out and about on the streets even though the sun had just come up. We sat together and watched the old men carrying papers and the young girls hurrying to work. I yawned again and looked over at John who was smiling.

"Remember the little *panaderia* in Cabo San Lucas?" he asked.

"Good palmiers, there," I said.

"They were better than these."

"Think so?"

"I don't know. When we were there I couldn't wait to get back to Paris for a real palmier. Now, I wish I were in Cabo again."

"You're just tired."

"Those were real palmiers down there."

I leaned back and yawned again but I was thinking about the little bakery in Mexico. The street in front was unpaved and there was always a lot of dust swirling in the wind eddies that rushed from around the corners of the buildings. It was a long way down the dirt road from where *Sun Flower* rested on her anchor line, but the coffee was strong and black and the palmiers were fresh. We kept walking back for more each morning and the *panadero* who owned the bakery smiled at us and gave us extra favors with our coffee. We liked being there just after dawn when the wind and the heat were down.

The little man showed us his ovens one morning. I stooped to see inside, but I brushed my head against a bare light bulb hanging on a wire from the ceiling. He apologized and held it aside for me. I could see it was the only light over the ovens and I wondered how many bulbs he had broken because there was no cover. I stood to leave while the baker held the light bulb away again. John ducked as well, but after I stepped out into the street, I wondered if I shouldn't find a hardware store to buy a proper light fixture. I looked in windows all the way back.

When we reached the inner harbor, you could see *Sun Flower* hunting about in a quiet morning breeze that came in from the southwest. We were ready to leave Cabo San Lucas then, but I remembered in a locker below I had a United States Navy-issue droplight that had a thick glass cover and a heavy aluminum frame. I sat in the cockpit thinking about the man and his dark ovens and how I needed to row ashore with the droplight and take it to the bakery. There is a lot to do when you are getting underway in a small boat and the hours wore on and in the end, we sailed away in the early afternoon to cross the Sea of Cortez. The droplight was still in the locker.

The wind died away to nothing that day. We motored into the fading light on a sea that looked like an alpine lake, so flat and blue and quiet we could see a shadow in the water a quarter-mile away. We were well offshore and there were no reefs to hit and so we slowed and crept closer to the shadow that loomed even larger as we approached. None of us aboard had ever seen a manta ray so big. We slid by in the quiet water and looked over the side onto the back of the ray that was more than twenty feet across, but being that close made me nervous for some reason. It didn't bother me that the ray was so big. I kept thinking of the shadow just below the surface and how much like a reef it seemed and I was relieved when we motored by the sleeping giant and into the evening shadows instead. I turned to look back toward the mountains above Cabo San Lucas, but we had gone too far and all you could see were the clouds. I was angry at myself because of the droplight.

I thought about the shadow on the water for a long time, even while sitting in the street café in Paris after the all-night train ride from Hendaye. The happy man in Mexico kept working in the shadows near the ovens turning out the palmiers that were better than the *patisseries* of Paris, and he treated us like old friends when we arrived each morning. The naked light bulb hung from the ceiling and I always imagined I would go back and give him the droplight so he could see the *bolillos* and the palmiers while they baked in his crude brick ovens. There were more shadows, though, and some reefs waiting in the dark. The droplight never made it out of that locker.

I looked at the people on the streets of Paris again. An old man with a push cart came along selling flowers and fruit and cups of juice and he stopped not far away to serve a little boy who held his left arm out straight with the coins in his fist. The little boy took his cup of juice and carried it away with two hands. The old man chuckled in the morning sun.

I liked watching the old man and his simple life. It made me sad to think of the bakery in Mexico without the droplight.

"You're right," I said. "Those were the real palmiers."

CHAPTER THIRTY-SEVEN

ringing...

The Big Sur country of California is a wild, hundred-mile run of lee shore where the cold northwest winds blow with unlimited fetch to crush the swells of the Pacific against the walls of the Coast Range. If you stand on the cliffs long enough to breathe in the wind, you can sense that wildness, like there is some tension in the air and a feeling of constant upheaval in the surf and in the mountains themselves, and there is Ancient here. It's like the Cambrian Period was just beginning.

A narrow roadway clings to the rugged boundary between the cliffs that loom over the booming Pacific and that frigid wind out of the northwest, and the mountains of the Ventana Wilderness that tower inland from the sea. I drove south on the Coast Highway out of Monterey just to see the Big Sur country again. I didn't stop until I came to an iron gate and the rutted path beyond it that winds down through Partington Canyon all the way to the beach far below the cliffs.

The gate to the canyon is easy to miss. I slowed around each bend and then finally parked along the highway only after I had driven past. I left the car and walked down into the gorge where there is a small stream that swishes its way to the sea. I stopped to listen for a moment, but beyond the stream there is a sheer rock wall with a tunnel that opens onto the isolated cove of Partington Landing. They say tanbark was carried down the canyon through the tunnel and loaded onto longboats in the old days, but the landing looks a lot more like a place for smuggling human cargo and the opium they brought from the Orient.

I was alone that day so I walked through the tunnel and stepped out into the bright sea air in the cove where the iron rings rust away in the salt spray and the old crane stands mute in the shelter of the cliffs. I could hear the wind over the noise of the surf as it whistled along the heights, but it was calm in the cove. I sat on a wide rock shelf and looked out over the kelp that drifted in and out with the rhythm of the surge. The wind shadow trapped the heat even though a high fog lingered over the coast. I leaned back and shut my eyes. I liked being close to the sea again. I didn't mind being alone and I let the sounds and the heavy salt air settle over me. I could have fallen asleep but the images came flooding back and I couldn't stop them.

The cuts on my feet were infected and my ankles were swollen because I'd walked a long way, so I took my shoes off and let my legs dangle over the sea wall. I'd been to see the Honduran officials, who were dull and distant and stupid, and while they droned on, the ringing in my ears was so loud I thought they must hear it. Through the open window came the sound of the jungle. I remember thinking how much it was like the ringing, and that it must be the jungle that made the sound in my head. There was no jungle on the reef, only the crashing of the waves and the terrible wrenching sound of my boat grinding to death on the coral. The officials were sweating as they spoke to me and I wanted to scream, but the only sounds came from the grating of the voices and from the jungle outside and from the flies buzzing in endless circles in the dingy little office. I sat alone on the quay, dull and vacant. The guards who passed by were armed with AK-47's. They carried them smartly when they knew I was looking. The ringing in my ears was deafening.

There was some ringing again, but I knew it was only the wind when I opened my eyes. The kelp in the shallows surged in and out, in and out, and I rubbed my eyes and stood and stretched my arms while the wind whistled over the cliffs. I was glad the landing was calm. I turned and entered the darkened cave and when I walked through and came to the stream on the other side, I dipped my hands in the icy water and splashed my arms and my face. I shivered at the cold and then trudged my way up the long fire trail back to my car.

Down in Lucia there is a café perched on the cliffs where the coffee is strong and black and hot and there are always travelers inside with faces turned rosy from the wind. I drove south from Partington Canyon and watched from my car as the surf pounded away at the base of the cliffs and the chaparral as it bent and shook in the cold wind that never stopped. East of the highway there were patches of blue where the fog had burned away, and the Ventana peaks were shining green in the sun. I thought there might be sun over the café. There weren't many cars on the highway and I drove the hills and curves in a hurry while the tape machine played music from the Sixties. The café in Lucia was a happy place. I didn't think there would be any ringing.

CHAPTER THIRTY-EIGHT

button faces...

After my years of sailing ended on the reef in Honduras, I did a lot of aimless wandering where nothing seemed to click. When an opportunity to work in Taiwan came up, I left behind what little I had and flew aboard a China Airlines 747 to Taipei and then caught another flight down to Kaohsiung where I started work as a foreman for Queen Long Marine.

Not many of the people in the yard spoke English, even in the front office. My time was spent trying to communicate in sign language and gestures and nods. I moved through the work days in an anonymous realm of noisy silence, but the boats took shape and the food was good in the lunch room. The guys who worked in teak and fiberglass and stainless steel teased me in fractured English. I convulsed them with unrecognizable Mandarin in return.

I moved into in an apartment building in the little town of Naosiung, a thirty minute taxi ride from the boat-yard. I needed to find a café in the village for morning coffee before the cab showed up. My day started early on, though, and instead of trying to walk down to the village at dawn I bought one of those hot water appliances that looked like a blender. It supplied hot or cold water instantly just by pressing a button. The man in the tiny open air market around the corner sold me a jar of Folger's instant coffee crystals. In the morning before my driver arrived, I made my coffee strong and cut it with sugar and canned milk and then sat outside on the balcony to watch the sunrise.

The horizon lightened slowly beyond the village but I could tell what kind of day was in store before the coffee was gone. By the time the taxi driver beeped his horn, the sky had usually taken on that whitewashed look, like the high Carolina sky of those summer days long gone, but sometimes there were angry clouds crowding over the hills to the east. There were slight gusts of wind that rattled the fronds on the fan palms and sent ripples across the water in the pool just below my balcony. The unsettled weather nearly always turned to storms at some point later in the day. The rain came so fast and so hard that the workers in the boatyard had to stop and watch with somber faces.

I drove out once on a Sunday morning with a friend to lay a running course for the local chapter of the Hash House Harriers. We wound up in a village twenty miles east of Kaohsiung where the town and the people of the rolling country made you think of the brush paintings that hung in the boatyard office. The hills were carved into small paddies of rice and there were flocks of ducks and chickens and tiny ponds for fish farming and the pigs wandered free along the roads where the water buffalo stood patiently, tied only by quarter-inch lines strung through the rings in their noses. The buildings in the village stood tired and rundown, though, lining the narrow canyon like Chung Yuan masks staring glumly at one another across the dirty street.

There were thunderheads billowing over the hills that day and gusts of wind that scattered bits of trash. When the sky turned black, we sought shelter in an open market near the center of the town. Jack and I stood together silently in the back and looked at the faces of the people who had come into the building from the street. Most of the faces we saw were expressionless, even when a little girl squatted and relieved herself in the gutter in front of the store. She stood and looked blankly at the people watching and then she walked up the hill and out of view. Her urine puddled briefly and then trickled in a pair of tiny yellow streams into a hole left behind by a missing chunk of asphalt.

The rain came quickly with the wind. We stood in the shadows while the roar built upon itself, so deep you felt an earthquake somewhere in your soul. At the peak of the downpour it seemed like the village had been moved to a spillway below a dam with the white water pouring down. We waited in the open market, consumed by the violence of the storm.

Lightning cracked nearby, highlighting the gray of the buildings and the black of the asphalt and casting shadows beneath the sills of the windows across the street. Water rushed down from the hill to the east to flood the gutters and pour over the sidewalks. Trash and the filth went sweeping along in the torrent that carried with it the urine from the little girl, all of it swirling together and cascading in the rapids caused by the missing chunks of asphalt.

People who were in the store with us watched the rain and listened to the wind and the thunder. They stood with their faces that said nothing, flushed red from the heat trapped in the stillness. There were flies in the market that flew in random patterns. Some of them landed on the faces and became dark specks on a dozen pale cheeks, and still the faces said nothing. Hard rain fell outside. Water roared and strobes flashed and shadows flickered, and I stood there wondering where the little girl went who had squatted in the gutter and left behind her puddle and trickle. I kept thinking it was all right to pee in the gutter because the rain would come and sweep the streets clean and there would be no more filth, only wet asphalt steaming in the sun after the thunderheads passed through. The villagers who stood with me in the store looked passively at the water tumbling down the gutters and still they said nothing.

The Starlight Bowl in San Diego is directly under the flight path of the jetliners landing at Lindbergh Field. It's an open amphitheater where the actors handle the jet noise by freezing momentarily while the planes pass overhead. When the noise falls to an acceptable level, the action in the play is resumed and the actors pick up their lines. As the wind dropped and the rain let up and the noise fell to an acceptable level, the peasants woke from their trance and left the market, one by one, hunching their shoulders and looking down at the wet asphalt while scurrying around the corner. The little man behind the market counter stood and watched the people go. The flies that left the faces stayed in the shelter of the store and landed on the glass front of the display case and on the balance tilting on the counter. I brushed one away that had been on my face and then I hunched my shoulders against the wind and the mist and hurried around the corner to find the car.

When the sun began to steam the road ahead of us, the glare was bright and the air was fresh to breathe. The afternoon was so pure it made me think of the office brush paintings again and little girls with button faces and rice paddies flooded deep from the storm. The stage for the peasants had come alive again. Like the actors who wait for the jets in San Diego, the peasants and the village people waited as the storm roared through. The wind and rain scrubbed their stage clean and carried away the somber faces leaving behind the laughter and chatter and puddles shining silver after the rain.

CHAPTER THIRTY-NINE

typhoon...

My apartment in Naosiung was on the third floor of a six-story building. On weekends the afternoon sun streaming through the windows sometimes made the living room too hot and muggy. Early evening clouds rolled in and blocked the sun, though, and then the rains fell, heavy and solid, and I pushed the patio doors open and let the wind blow through. The smell of the rain was fresh and cool and I sat outside and breathed and listened to the sounds of the water running off in the gutters.

When the rain let up briefly, I could hear behind me the chirping of the geckoes that climbed the walls and hung from the ceiling in my apartment. They sounded like sparrows in the spring, but the rain would fall heavily again and all I could hear then was the roar of the drops when they hit the fronds of the palm trees and the concrete of the sidewalk and the stagnant water of the swimming pool. The sound was so loud it made me think of those freight trains in the night, but always near sunset the storms moved beyond the village to rage against the hills to the east. The wind and the noise slowed to a whisper while the sound of the geckoes chirping in the apartment returned even louder than before.

After I got to work one day, the manager of the boatyard told me they were shutting down for three days. I thought the date must be some kind of Chinese holiday, but the workmen in the yard were nailing boards over the office windows and all the loose equipment had been moved inside. One of the men who worked in fiberglass came over to me and said, "Typhoon." I had nothing to do then so I called for a cab using the only Mandarin I knew. The driver was new and he took a long time to get me back to my village. Along the way there were all of the street people gathering up what little they owned and cramming it into their shacks. I wondered what good it would to do.

Not far from my apartment building there was a family who lived in a twenty-foot steel shipping container. The father had used an acetylene torch to cut doorways into the sides and then he added rooms out of scrap wood and cardboard and it made a crude but workable home that stood by itself at the edge of a barren, ten-acre field. His family had accumulated a great wealth of industrial discards. All of these goods were stacked neatly in the field behind the container and behind the add-on rooms and in the dirt near the entryway. When we drove by in the taxi before the storm, I could see three tiny kids sitting in the doorway looking up to the sky. Their laughing, moon-shaped faces made me think of those little blubbery dots that bounce over the song lyrics on "Sesame Street".

My own apartment nearby was bombproof. The driver left me in the middle of Niaosung so I could stock up on beer and other essentials. When my shopping was done, I carried my purchases back and set up camp to wait out the coming storm. I had been in a few hurricanes while living in the coast country of South Carolina and so I sat on my patio and watched for the changes in the sky and listened to the geckoes chirping at my back. The Taiwan beer from the village store came in sixteen-ounce bottles and there was a stack of them in the refrigerator. In the calm time when the sky turned yellow and the heat was trapped near the ground, I stayed outside and drank several of them. It seemed like the wind and the rain would never come.

By sunset the typhoon had roared in with such fury the village below me looked like the sound stage for an old black-and-white film from the thirties where the rain blows in horizontal sheets and the actors stagger like drunks against the wind. When I had to move inside with the geckoes, I shut the sliding glass door with a bang. It didn't take long for the lights to go out. I had to burn candles to see and I sat on my couch and listened to the roaring and the screaming outside of a hundred freight trains. The geckoes crept along the walls while the flickering light from the open flames made their shadows jump and quiver. The lizards seemed bigger than life. I ignored the iguana-sized shadows and sat on the couch and drank the warming beer and then collapsed into bed around midnight.

After daybreak, the sound of the wind and the rain had subsided. I put on a foul weather jacket and climbed the stairs to the roof of the building to see what the typhoon had done. Most of the village structures were still there. I could see through the drizzle the village people trying to spread those sheets of blue plastic over the yawning holes where the roofs used to be. To the west and south were the rice paddies, but the tall and green stalks of the new harvest were flooded and there was nothing but water for miles. It was like looking out to sea again, but the colors weren't right. All I could think of were toilets that overflowed and sewers backed up in the rain.

After the floods drained away and I returned to work, my regular cab driver drove me through Fengshan to check on the family that lived in the shipping container. When we turned the corner, I only wanted to see those kids with the moon faces again bouncing around outside and playing in the puddles left behind by the storm. The steel container sat in the mud by itself and there were no added-on rooms anywhere, no stacks of discards placed neatly behind, no kids in the front looking to the sky. I only saw the container standing alone with its gaping doorways staring out to nowhere and scattered bits of junk reaching from the field like the legs of so many geckoes drowned in the muck. I asked the cab driver what had happened to the family. He shook his head and looked out of the window to the field of mud.

"Typhoon," he said.

CHAPTER FORTY

the binjo...

I could see the pool from the patio outside of my apartment in Naosiung. No one took much care of it so the water didn't look very often like the Caribbean. Someone flushed and cleared the filters and fooled with the chemicals and the pool came clean and turned blue like the water in the fountain in front of the coffeehouse here, but then it went neglected and the algae began to bloom all over again. The Europeans who lived in the apartments swam in the pool when the water was clear but as it began to sour, only a few Taiwanese would go near it. As the days went by, the pool exploded with algae until it got so dark that not even the hardy Taiwanese could stand it and then the water turned nearly as black as the water in the drainage ditch across the street. Summer heat can be brutal in Taiwan, though. The humidity can match the temperature of the air and in the south of the island where it's tropical, you can't walk from your apartment to your car without soaking yourself in sweat. After the pool turned black and stagnant for a few days, someone cleaned the filters and fooled with the chemicals and the cycle started all over again.

The Europeans and Americans and some of the Taiwanese locals enjoyed running with the Hash House Harriers on Sunday mornings. After scrambling over mountains and through rivers and rice paddies and jungle and suffering through the heat and the humidity of southern Taiwan, you were rewarded at the finish line with some chanting and singing of raunchy cheers and with cases of

Taiwan beer on ice. No one ever figured out whether the Hash was a running club or a drinking club, but we ran through the countryside and drank gallons of beer in the summer heat. When the runs were finished, many of the people retired to the pool in Naosiung when it was clean and clear. Most of us retired to the bar in the Kingdom Hotel downtown to drink more beer.

Near the village of Jenwu, a low wooden bridge with no rails crossed over a wide drainage ditch the locals called a *binjo*. The bridge itself was rough and splintery and not very substantial. When Eddie ran toward it he tried to jump the putrid creek instead. He came up a little short. After he crawled out of the binjo, he was covered with brown scum and slime. I thought he would die of hepatitis or something even worse because the water was the color of bittersweet chocolate and just as thick. The drainage in the binjo was so slow the chocolate was nearly stagnant. We could smell Eddie from a hundred yards away because the ditch was filled with the waste of the animals and of the people of the village.

After our run that day, the Hash House Harriers gathered in the Kingdom Hotel again. Eddie took a seat at the end of the bar where we watched him wash and scrub his insides with bottles of Taiwan beer. He smiled and talked and you couldn't tell he had been in the binjo unless you noticed the brown stains on his shirt and on his running shorts, but then he shuddered and trembled and guzzled his beer to keep from throwing up. All of the runners who sat at the bar tried to laugh at Eddie and make him see the humor, but there was no humor in it for him. He only grimaced and turned away. In no time at all he was too drunk to sit upright on his barstool. We loaded him into a taxi and I told the driver to take us to Niaosung. Eddie only made it about half-way before our driver had to stop to let him throw up at the side of the road. He only just made it, though. The driver was so afraid of accidents in his car that all I had to do was clear my throat and the taxi would dive for the side of the road in an explosion of dust and gravel and chicken feathers.

Eddie never got sick from the water in the binjo. He only got sick from the beer. The sounds of his retching went echoing down the hallways of our apartment building while I helped him to his room. I thought there might have been a problem with some of the Taiwanese residents because of all the noise, but the doors stayed closed. Eddie retched and stumbled and lurched toward his apartment, scattering the geckoes that clung to the walls of the hallway.

I left Eddie fully clothed and standing half asleep in his shower stall and then took the elevator up to the third floor. When the automatic door opened, I looked down the hall toward my apartment. Hundreds of geckoes hung from the walls and from the ceiling. The air was still and hot and muggy and it wouldn't have been much better in my room. It was only nine o'clock at night so I punched the button in the elevator for the ground floor. The Hash House Harriers were still drinking beer down at the Kingdom Hotel and it was too hot for me not to do the same.

There weren't many taxis in our village at night. I stood near the entrance to the parking lot waiting for headlights. Some minutes went by and there was nothing on the road that curled in from the east. I stood alone in the dark and in the heat and in the smell that drifted up from the binjo nearby. Beads of sweat grew into trickles that crawled down my back. The smell of the binjo across the street was full and ripe and reeking and it seemed like my shirt had picked up the odor. I kept rubbing my nose with the back of my hand, but the smell wouldn't go away and it was really hot and still on the side of the road. After a few more minutes, I started to taste some of the beer from the Kingdom Hotel and still there were no headlights to pierce the night. I thought about all of those brown stains on Eddie's shirt and then I wondered how much of the brown stuff had smeared onto me when I helped him out of the ditch and that maybe it wasn't the binjo that reeked in the night but me. I didn't wait any longer for a taxi.

When I got to the third floor of the apartment building, there were all the geckoes hanging upside down in the heat of the hallway, chirping like caged birds. I liked the geckoes and I didn't want to disturb them, but I couldn't get to the shower fast enough. I stumbled down the hall while the geckoes scattered like fifth-graders running from some fat kid passing gas. The air in my apartment was thick and smothering and I could smell the binjo again. I walked out onto the balcony to breathe. I was shuddering from the odor and from the brown stains on my shirt so I drank another beer and stood in the cold water of the shower. It took a long time before the shuddering went away.

For several weeks, Eddie couldn't get used to the smells of Taiwan after he fell into the binjo. He thought he might leave his engineering position and return to Iowa. He was one of the good guys, though. He knew what it meant to live in a sub-tropical island with sixty-two mountains over 10,000 feet high and department stores that rival Herrod's in London and peasants who herd ducks by the hundreds. After a month, he was back to running over bridges and jumping over more binjos and drinking beer with the rest of us. He never asked for a transfer back to Iowa.

I would have stayed myself, but then there was the Caribbean and that reef off Honduras. I felt like there was some unfinished business down there. I left Taiwan behind thinking I would make my way back to the islands. My world in San Diego was changing, though, and the Caribbean didn't seem so important to me once I settled in again. A lot of years disappeared. I never made it back to the reef to find the bones of *Sun Flower*. These days I'm happy to sit by myself in front of the Hot Java Café and enjoy a cup of Old Foglifter. I'm always staring at the fountain outside where the water dances in the sun and throws diamonds against the wall. The pool beneath is so neon blue you can imagine Dominica or St. Vincent and the Grenadines, or even the reefs off Honduras if you shut your eyes and let yourself drift away.

Sometimes I listen to the splashing of the fountain and think only about Taiwan. I loved being in that country. I always thought I would have the financial means to return to the southern part of the island where the summer storms and the stop-action lives of the village people, even the binjo that trickled across the street from my apartment, seem as alive to me now as they did twenty-five years ago.

All the towns and villages in Taiwan are laced with those binjos. Even the Love River that flows through Kaohsiung looks like dark chocolate. It's a long way to the neon Caribbean. There is something exotic, though, about living in a country where you aren't able to read the newspapers or the street signs or order from a menu or ask for directions from the little man selling *wanchu* in the night market down the street. Living in a place like that allows you to be separate from the people, as if you were the exotic one. In those brief instances when you do understand their world, it's like discovering for the first time there could be such a thing as a bridge that crosses an impossible cultural gap. Suddenly you become part of their country, part of the mountains and the sea and part of the rivers of chocolate and the thick, slow-moving binjos, and part of the peasants herding all those ducks in the villages where the water buffalo wade silently in the chessboard paddies of rice.

CHAPTER FORTY-ONE

Ranger Rick...

I was thinking about my work in Taiwan a few minutes ago when I realized how long ago that was, and not just in terms of time. There were periods in my life when I thought a house and a picket fence were death knells, yet here I sit in a suburban coffeehouse where you can hear the sounds of the freeway each time the door is opened. I have to teach algebra tomorrow. The thirty-year mortgage where I live with my family is up the hill and around the corner from the shopping center. On days like this, it seems there is nothing left of the Caribbean or Key West or Mexico, nothing left of Europe or Newfoundland or the Orient or the endless seas of times past, nothing left of the catfish or the Waterway or the Carolina Lowcountry. I don't know how I got here, but at least I can smile about it. Unlike a few people I know, I had some time to spend on life.

A man is sitting with a woman at a table near me in the front of the coffeehouse. They're talking about his relationship with God and about the relationship he has with his wife. The woman doesn't agree with what the man is saying and their voices sometimes get loud. It's hard to sit here and ignore them. He believes in the Bible and he's telling the woman that he tries to live by the word of God and that he's working hard to save the marriage. She told him she wishes he were dead. I don't like sitting so close to them because of what they are saying to each other and because the woman is wearing too much perfume. The fresh air vent in the ceiling near the front is blowing the smell of

NutraSweet flowers over to my table. All I can think of is an overweight lady I heard once singing "I Feel Pretty" so far off-key it made my rear-end itch. There's another man sitting across from me who is absorbed in the best-seller AGELESS BODY, TIMELESS MIND. I wish he would share some of those philosophies with the couple arguing by the window. There's a lot more to life than struggling with God and with a wife who is trying to kill you by wearing too much perfume.

My son Casey and I went to get haircuts last night. While we waited in the cushioned chairs at the front of the salon, a woman came in and took a seat near me opposite the magazine rack. She wore her black hair bobbed and an oversized sweatshirt inside-out over a blue turtleneck and a pair of knee-sprung jeans that were too long in the crotch. I watched her for a minute. I was thinking of the people who vacation on the Russian River when I noticed she had taken from her backpack a hardbound volume of Randy Shilts' book, AND THE BAND PLAYED ON. She leaned forward slightly to open the book and it was like watching a child open a first volume of Mother Goose rhymes, but then she began to read. In a few minutes she came to a part of the book where she suddenly took a big, slow breath and then she put the book down for a moment and looked away. You could tell she was struggling with her composure. She kept taking those deep breaths and looking through the window, but then she turned and began to read again the book by Randy Shilts. I watched her in silence while she held this man's monument in her hands.

I'm not able to visit the wall in Washington, D. C. There are too many names and too many tears and all of those flowers. There is so much we could have changed to make the wall unnecessary, but I'm glad it's there. Maybe when I'm an old man I'll be able to sit on a bench in the sun near the wall and remember the boys I knew without being overwhelmed with emotion. For now, that's not possible.

While I watched the woman in the haircut salon reading her thick red book, I was reminded again of the wall and of the names engraved on its face. I thought of the artists and writers and dancers and people of music we have lost to AIDS. There are no walls for them, but at least we have the books and the music and the paintings. When you think about all the others who have died, you wonder who will remember at all.

I had a friend at work who became ill and had to ask for some time off. When he was gone for a few days, no one worried. After he had been gone for a few weeks, a lot of us became concerned. He didn't come back until the following school term and when I saw how thin he was, I knew. He asked me if I would join him for lunch and in the privacy of his car, he told me his secret. He told me there were a few others at work who knew but that he wished it to remain a secret until he was forced to quit because of his health, and then he planned to tell everyone. When that day came, a meeting was called. We all sat in the chairs in the library while my friend stood tall in front of us and told us he was dying of AIDS, that soon he would be leaving.

When someone tells you he is dying, it's hard not to be awkward. It's hard not say things corny and stupid and so you don't say much at all. You only wring your hands and look out of the windows and try not to cry. The one who is dying will just stand there.

Over the next few months I watched my friend go. I didn't say much. I helped him put together his condominium so that his brother would have an easier time with the real estate people. I helped him move a few things out of his life to make it better for him when the time came. Mostly, I was just there. We drank a lot of coffee and talked about kids we had taught. We talked about places we had been and people we knew. When there was no conversation, we sat and looked through the windows at the trees and at the sky and at the low clouds, gray and thin, that drifted in from the sea to the west.

There were no headlines in the end. My friend died of complications brought about by the AIDS virus, and that was all. There is no wall to touch. So many of the others have the books and the music and the paintings, and for the actors and dancers the film, but for my friend who was a teacher and who spent his short life only giving, there is nothing. On the day of his service, I couldn't force myself to go. I stayed home while another friend of mine came over and in a cold, driving rain we transplanted a ten-year-old palm in the mud of my back yard. It stands tall in the corner near the fence. I can't look at it without thinking of Rick. It might be the only wall he has.

The two people who have been arguing in the front of the coffeehouse are leaving. Neither of them is smiling and it makes me think nothing was resolved. They are in a bad situation, but it's not easy for me to feel compassion, only pity. Life is short and fleeting and there are no guarantees how or when it will end. It seems extraordinarily frivolous to sit in a suburban coffeehouse and argue when all we have to life is a little bit of time. Those two are lucky to have any left to spend. I wish my friend Rick had some.

CHAPTER FORTY-TWO

crappy - all you can eat...

The twilight sky has taken on a color that reminds me of the sea that drifts across the sand on the shallow banks of the Exumas. I don't normally drink coffee outside because of the faint smell of cigarettes, but the Bahamian color of the sky is intense. It's also quiet tonight. I wouldn't have known a hawk was near except the haunting sound of its cry carried all the way into the shopping center. The blackbirds nearby are hopping about searching for pastry crumbs, whistling so softly you think of air escaping from punctured tires. I'm glad the Starbucks is nearly empty.

During a road trip through the South a few years ago, my driving partner and I stopped on a bridge over a creek in Alabama just to see. The morning was quiet with very little traffic behind us on the blacktop. We leaned against the rail and listened to the sounds of the creek bottoms. The heat and the stillness and the heavy air that settled over the roadway made us sweat in no time, but the exotic sounds of the chirps and the buzzes and the faraway screeches of the hawks in the distance held our attention. We stood on the bridge while the sweat soaked the backs of our shirts and the faint smell of road tar drifted about in the stagnant air, but we stayed for a long time just to listen to the noise of the cicadas and the frogs and the mockingbirds, and to the slow gurgle of the creek.

Back in the car with the windows rolled open, we drove another five miles down the highway and passed a sign that announced, "CRAPPY – All You Can Eat - $2.95".

We pulled into a gravel parking lot in front of a roadside restaurant that had been painted green once a long time ago and then stepped inside for lunch. A nice looking lady who cooked and waited tables and washed dishes stood alone behind the counter. She told us we could serve ourselves from the heated trays of fried fish and from the cooler that held a small mountain of salad. We piled our plates high and ate deep-fat-fried crappy and homemade coleslaw and when we were finished, we went back for seconds and took the last of the crappy from the server and a good part of the salad. We had just started in again when a man walked in and took a seat at a table near the counter.

"This place is filthy," he announced to the lady. "You better sweep. You need to refill the catsup bottles on the tables, too, but bring these guys some water before you start."

To keep him from ordering the woman around, Mark put down his fork.

"Where did you get the crappy?" he asked.

The angry man's demeanor changed briefly.

"Caught the fish myself from the river a half-mile east of here," he said. "Just used a cane pole and a feathered jig in the shadows beneath the highway bridge. The crappy hide where the water's dark."

He glanced over to see there were no crappy left in the warmer. He turned to the woman again.

"You better fry some more crappy for me. I ain't ate since this morning."

"Honey, there ain't no more crappy," she said softly.

This made the man angry again and he stomped behind the counter to check for himself.

"How could you be so stupid?" he yelled. "Why didn't you save some crappy for me?"

It was hot in the restaurant and so we left some money for the ill-tempered man and his poor wife and we drove away in the heat and the stillness toward the river. When we reached the bridge, we stopped and looked over the side into the shadows where the dark water hid the crappy. I remember thinking how good it was to be free of the man who yelled at his wife and to be able to stand on the bridge and hear again the sounds that came up from the river bottoms. The heat was intense and the river bottoms were alive. I thought one day someone might come who would take the woman away from the crappy man who did all the shouting in the restaurant and bring her quietly to the bridge, just to let her listen.

◊ CHAPTER FORTY-THREE ◊

heaven didn't work...

I stood in line for coffee a few minutes ago when I turned around to see a kid walk through the door wearing skin-tight Levi's and a belt buckle the size of a small manhole cover. He took such long strides in his boots that he was behind me before I turned back to the counter. I smiled and nodded. The kid tipped his 100X Stetson without saying anything. After I got my coffee, I took a seat next to the window. The cowboy got his coffee to go and was already out on the sidewalk headed for an F-150 waiting over by the travel agency. The kid ducked his head when he got into the truck. I watched while he backed out of the slot and drove away through the parking lot. All I saw then were the Montana plates and an empty gun rack and that big black Stetson in the window.

During those years of war in Vietnam when I went through a tour of duty in the Navy, my best friend was a kid from Montana who worked with me in the Training Aids Division at North Island. He was the purist athlete I have ever known. A good part of his time was taken up with golf and tennis and football and he was the starting point guard on the championship basketball team. When the base held a table tennis tournament, Jeff went through the rounds undefeated and won the trophy. He was also the shortstop on the base softball team. He confided to me once that hitting a fast-pitch softball was the one part of sports that escaped him, but no one ever played with more intensity and with more love for the game than that kid from Montana.

Jeff was married at the time to a girl he had met in high school who was from the old mining town of Butte. They talked endlessly of going back to Montana to settle down so Jeff could return to Montana State in Bozeman to earn his degree in the wildlife management program. This was his dream. Janie and I heard stories of aspen trees and mountain cabins and of rivers sweeping through and meadows where the antelope would move in to graze just before sundown. They told us of hunting and fishing in the mountains of the back country and when they invited us to dinner, we enjoyed elk steaks and venison stew and drank beer until the early morning hours. We talked forever about the Montana that always seemed like heaven to them.

A short time before Jeff was to be released from active duty, Janie and I left for a two-week vacation to Montana to see for ourselves what it might be like to live there. Jeff had told us to look up his father in Boulder. When we did, we were invited to spend a few days in the Montana high country where we helped build the foundation for a cabin out of local stone with mortar we mixed by hand in a watering trough. While we were there, the elk crept from the shadows of the aspens to feed near the brook tumbling softly nearby, and the sounds of the hawks flying over the hills to the west pierced the evening silence. In the middle of the aspens up on that ridge, I knew absolutely that Jeff was right. Montana seemed like heaven to us, too.

Within a few days of our return to San Diego, we helped Jeff and his wife load their belongings onto a home-built trailer. We said our good-bye's while they drove away, heading north into the morning. There were no tears because all of us thought our separation might only be temporary. I wasn't due to be released from the Navy for another year, but Montana wasn't going anywhere and it still seemed like heaven. In the end, though, there was a boat to build and a life of sailing ahead and we never saw our Montana friends again. We exchanged letters at first and a few Christmas cards. We knew that Jeff was in the graduate program at Montana State and that heaven for them was just the way they thought it would be, but Janie and I left for our own kind of heaven. The letters slowed and then trickled and then stopped. When you are anchored in a quiet cove down in the tropics where the air at sundown is only baby breath and the long lines of shadow you see are cast by the coconut palms leaning over the white sand on the beach, it's hard to remember the aspen trees of Montana or the eagles beyond the hills or the elk that came down to feed in the shadows of the late afternoon. Forty years went by in a heartbeat.

I went back to Montana with my family a few years ago. I wondered if heaven was still up there. I didn't have much of an itinerary, just that at some point I might wind up in Boulder again and that I might try to find Jeff's father who I remembered as being one of the world's great human beings. When I found the old man, it took a while for him to recognize his visitor, the one who had mixed all that mortar so long ago. He invited my family into his home and then he sat stiffly on the couch and told with a lot of emotional restraint the story of his son.

The masters program at Montana State required a thesis before Jeff could earn his graduate degree. He set out to document the movements and the breeding habits of a herd of bighorn sheep he discovered in the mountains above Bozeman. Milling about among the ewes was the largest bighorn ram he had ever seen. Jeff worked for a long time tracking the herd as it wandered through the remote canyons and up the towering cliffs, recording his observations as he went, but the ram was so big and the horns it carried were so magnificent that Jeff began to wonder if the sheep would qualify for a Boone and Crockett award.

Jeff began to toy with the idea of hunting the ram down and killing it after his thesis was complete, but a slight moral dilemma stood in the way. His research as a future conservationist and as a future employee of the state led him to the ram, not his hunting prowess. During the last stages of his field work, the temptation for killing the ram began to grow. Out of concern for his work future, he asked some of his faculty advisors what the consequences would be if he returned to the herd and shot the ram. The advisors were alarmed, of course, but they were powerless to stop him. They voiced their concerns over public reaction to the killing of such an animal, especially in regard to his status as a student in the fish and wildlife program. They advised him not to go through with the hunt.

Montana was heaven to Jeff, though. He was born and raised during that time when you still shot an elk and you still shot your deer and your antelope and you packed your freezer with meat for the winter and that's how you lived. You took the heads with the antlers and the horns to the taxidermist and when they came back, you hung your trophies in the living room. Through all of this there wasn't much regard for conservation or wildlife management because the elk and the deer and the antelope had always been in Montana and it seemed to a lot of people they always would be. Jeff hunted the big ram down and shot it.

A moment of anguish crossed the face of Jeff's father who sat with dignity on the couch in Montana. I didn't know what to say. In those awkward moments of silence, I could only look around the living room at the mounted heads of elk and deer and antelope and at a lone bighorn ram staring out to nowhere from its dusty place on the wall.

Montana made it very clear to Jeff he would never work for the state in any capacity. Since there were no other avenues of employment for someone trained in fish and wildlife management, there wasn't much for Jeff to do. He worked in construction for a time with one of his brothers and he did some surveying and some wrangling, but Jeff began slowly to recede from Montana because there wasn't much left of his heaven. There was nothing left of his dream.

Jeff's father wrote on a slip of paper the address in British Columbia where Jeff and his wife and their two sons had moved. The best athlete I have ever known is a cowboy now. He does some work for an oil company up near Benson Lake and they have a small piece of property where they board and raise horses. Life seems to have settled in for them in a new kind of heaven.

I've been carrying that little piece of paper around with me ever since my visit to Montana. I stumble across it whenever I go through the ATM slips that clog the pockets of my wallet. You can see on it the shaky scrawl of an old man. When I read the address, I think about Jeff and his family and about all the ranchers in British Columbia. Sometimes I think about drive-by shootings and about some drugged out guys in ski masks holding up a mini-mart, and I think about babies killed by teenaged mommies who don't want them. I'm just trying to keep things in perspective. Killing a bighorn ram in Montana couldn't have been so bad a thing to do. For Jeff who was the best of friends, the shot he took changed his life. Suddenly, there was no more heaven.

I haven't written to Jeff. I can't tell you why. Maybe it's because I would rather have our friendship stay as it was during the Sixties. Maybe it's because I saw the pain in his father's eyes when that old man decided not to hide what his son had done but chose instead to be honest. Maybe I'm just disappointed that heaven didn't work for either one of us. British Columbia is more beautiful than ten Montana's. I only hope there has been some peace for Jeff up there.

CHAPTER FORTY-FOUR

prance in silence...

I sat here for a long time staring at a big mural that sweeps across the ceiling of the coffeehouse. I didn't know there was any design to the painting, only that it looked like a lot of random brush strokes in browns and grays, but there was a moment where, suddenly, I saw horses prancing and manes blowing in the wind.

The reaction I had made me think of a teacher I once knew who tried to introduce her fourth-graders to the concept of photosynthesis. She gave to each of the students a bit of clay and instructed them to create six molecules of CO_2 and six molecules of H_2O by rolling the clay into marble-sized balls and connecting them with toothpicks. When the children finished, she explained to them how green plants use the energy of sunlight to convert these molecules into a simple sugar the plants use for food. She demonstrated how to take the water and carbon dioxide models apart and to redistribute them into a larger sugar molecule she labeled $C_6H_{12}O_{6,}$ but when she was done, she had all these leftover oxygen balls on her desk. You could see the confusion in her face. While she sat there staring at the toothpicks and at the clay on the table, some of the kids began to roll their clay into snakes and other kids smashed the clay and made cookies. One of them threw a clay ball against the wall where it flattened and fell down into the rug. The teacher sat puzzling over her project, but there came a time when it looked like an invisible two-by-four swept across the room to hit the teacher full in the face. She fell back and said, "Oh...

Oh...I just learned something."

The mural on the ceiling was all scribbles until I saw the horses. I felt stupid like that teacher who took so long to understand how oxygen is released in photosynthesis. Now the horses above me look real. When you sit here with your coffee, there is the image over your head of mustangs tossing their heads and stomping the earth and dust is billowing through the branches of the mesquite. The ceiling looks like a holding pen fresh after a spring round-up.

Horses are in my blood. As a kid I dreamed one day I would be wealthy enough to own a ranch where I could have horses in the fields just so I could watch them run. I wasn't much of a rider and I never wanted anything from the horses other than for them to be free and to run with the wind. In my dreams the horses were always healthy and groomed and their coats would shine in the sun and I would sit on the fence just to watch. I still think about that.

Back in the early fifties, George Air Force Base was a sprawling oasis in the middle of the high desert east of Los Angeles. Our house was on the southern perimeter of the base where, just across the street, the desert fell away for miles. The military boundaries weren't fenced so we hiked the hills and the buttes and the arroyos all the way to the Mojave River that trickled its way down from the snow pack in the mountains to the west. All of the kids in my neighborhood were constantly roaming the desert around the base. On weekends we sometimes made it to the river where we played in the cold water and ran through the trees that grew in the soft sand in the bottoms.

On one of these Saturdays near the river we found a horse. We only thought the horse was asleep in the shade and we were sneaking up on it when the wind freshened and then all of us were hit full in the face with a vile and putrid two-by-four that stopped us dead in our tracks. The horse still only looked to be sleeping and so we crept around to the windward side and approached it, leading with our noses flared wide like nervous colts. When we stood over the horse, flies buzzed so loud there were no sounds to the river nearby. They were clinging to the sockets of the eyes and crawling into the mouth and into the nose. The flies were so thick you couldn't see anything but the swarming. One of the guys pulled hard on a hind leg. Part of it tore away from the carcass to reveal a convulsing mass of bloated maggots that pulsed and writhed and roiled, pale against the rank interior of the horse. A cloud of gas hissed from the body cavity and poured over us like vaporous feces. We shuddered and ran from the stench and from the flies and from the maggots that seethed and churned and gorged themselves on the flesh of the rotting horse.

Boys of eight and nine don't have the option of throwing up in front of each other. We ran from the horse toward the river and giggled out of nervousness and tried to make crude jokes about eating maggots and horsemeat for dinner. We kept running through the river bottoms because if we had stopped, some of us would have been sick. The water in the river shocked us it was so cold. When we finally slowed, one of the guys said, "God...I never saw anything like that before," and all of us in the river trudged along in silence thinking about those maggots and about the horse that lay dead in the shade.

I was arrested in central Oregon once after I made a bad decision. Before I was to go to trial, some friends posted bail and I returned to Fort Lauderdale to gather my wits. I knew my time at the boat company was over. The guys who were taking my job had already cleaned out my gear and there was nothing for me to do but go through the motions. An Englishman who worked for me and who had lost his upper centrals in a bar fight now was to be the manager. As a gesture of friendship, he asked if I wanted to go with him to Hialeah for the running of the Flamingo Stakes. There were no reasons not to go so I called a lady friend who hung around the Cat's Meow down near the 17th Street causeway. The three of us drove to Miami in the boat company van for a day at the races. I was mildly in love with the bar girl, but she didn't know I was out on bail. I didn't want to hurt her and so we only just flirted and teased and bet our money on the horses that went by in rolling earthquakes like the horses that galloped across the fields of my boyhood dreams.

Just before the Flamingo Stakes was to be run, the bar girl went to the window to place our bets. While she stood in line, an old man in front of her turned to chat her up. By the time they got to the window, he was mildly in love and gave her a hug and told her to bet on a horse called Tap Shoes. She squealed with delight and put all her money on the old man's horse. When she came back, she told me what had happened. I put my money on a little horse out of Foolish Pleasure because I thought it could run, but I was only just guessing. I wondered why she didn't come back to ask if I wanted to change my bet.

When the starting gates opened, there was a dull roar from the crowd and we all surged to the rail to get a better view of the horses. When they rounded the turn, we were swept up in the roaring and the screaming from the grandstands and we began to scream as well. You could feel the earthquakes again from the pounding of the hooves and it was like standing in front of an avalanche. We were jumping and screaming because the horse in the lead was the little Foolish Pleasure colt. Running on his left flank was Tap Shoes and the Englishman's gray was on the outside moving up fast. The guy kept screaming in Cockney through his toothless mouth, "Run, run, you shilly bashtard!"

The horses thundered by and the earth shook and the mud from the track flew into the people who jammed the rail. The roar of the crowd swept down the stretch with the horses and at the finish everyone was screaming and screaming again and then the noise of the crowd ebbed for a moment, like someone feathering the throttle on an Allison V-12 while they held their breath and watched the lights on the tote board. When the winners were flashed to the crowd, the roaring from the grandstand shook the late afternoon like the Allison in full bore. My lady friend from the Cat's Meow screamed with them and jumped into my arms and she kept screaming in my ear, "I won...I won...I freakin' won," and then she dropped from my arms and ran laughing up the ramp to find the drooling old man.

I stood by the rail and watched the crowd milling about while the Englishman hopped around looking for the bar girl who hadn't come back from the windows. A slight breeze came in gusts and picked the losers' tickets off the ground and swirled them around like bits of lime in a glass of gin. In the quiet time after the horses had gone through, I couldn't think of much other than the cold jail cell back in Oregon. I didn't want the horses to stop running, only to keep circling the track like the flamingoes shocked from the infield lake for the television cameras. I wished the bar girl hadn't gone away.

I was driving west on a highway outside of Valdosta, Georgia, when I heard on the radio that John Lennon had been shot dead. I didn't want to attract attention but I pulled to the side of the road and stopped, breathing deeply with my head tilted back so the tears wouldn't spill. A man was driving a mule team along the furrows in a field next to the highway. The dust behind him swirled in long red ringlets that drifted clear to the line of trees in the distance. The dust hung in the air like that moment in time before an accident when everything unravels around you in slow motion and your vision is as clear as spring water.

I remembered an old black man driving a mule and plow in a field near my house when I was seven. An older boy in the neighborhood who had some kind of mental challenge came by to see me so I ran with him out into the field to help the man and the mule. We tagged along behind in the dust and the old man tipped his hat and told us he was going to plant corn in the deep furrows that ran straight as far as I could see. The man clucked to the mule and shifted the straps that crossed his back and the mule pulled the plow straight toward the afternoon sun. My friend began to cluck to the mule and the old man said, "Whoa," and my friend said, "Whoa," and the mule stopped. My friend said to me, "Hey...I can make this old horse stop," and every time the old man clucked or said whoa, my friend did the same. He turned to me and said, "See what I mean?" and he laughed at me like a little Frankie Fontaine.

There was so much dust out in the field, though, and you could see the mule and the old man were good and patient and were only ignoring my friend. The field was being plowed straight and true without his help. I stopped to watch the three of them plow the furrows while the dust swirled all around me in those lacy red ringlets. When I left them behind to walk back home, I turned to look once more. The old black man was still behind the mule that plowed the furrows and my friend, who tried to comprehend the world through his own twisted ringlets, scampered along beside the mule, clucking and yelling whoa. He looked in the distance like a horsefly that couldn't be brushed away.

There were ringlets of dust beside the highway from the mule team in the fields near Valdosta. The night before, a twisted kid hung around the Dakota Apartments screaming, "Whoa!" John Lennon stopped to see.

The girl behind the counter came over to talk for a moment. I told her about the mural on the ceiling and about the horses on my legal pad. She smiled at me and then looked away through the windows of the coffeehouse. She was a cowgirl once and she told me of a year in her life when she followed the rodeo circuit. While she talked, the meadows of green and the streams that tumbled clear and cold from the hills flashed in her eyes. The horses were still in her blood.

The same horses prance in the mural across the ceiling of the coffeehouse, milling about and tossing their heads like that childhood dream when the horses ran like thunder across the meadow and I sat on the top rail of the fence to watch the streaming of the manes and the tails and the flight of the hooves and the flash of the coats in the sunlight. There's nothing outside to see but asphalt. The coffee girl looked back at me and then she took a big breath and walked behind the counter. Horses stay in your blood, you know, even if they only prance in silence across the ceiling of a coffeehouse.

CHAPTER FORTY-FIVE

the good Bustelo...

Behind the counter tonight a young girl is busy wiping the Formica clean. I caught myself staring at her while she worked. I didn't mean to be rude. Her ponytail is tied with a ribbon the color of the sodium vapor lights in the parking lot and when she moved, her hair fell across her shoulders. She shook her head to clear it away and then she glanced my way to smile and talk for a bit. Her voice was soft and musical and it made me think of high school girls in Southern gowns. I didn't say very much, but after a moment I asked her if Starbucks had any Bustelo. She giggled and shook her head. I don't think she knew what it was. I don't even know why I asked her except that she reminded me of someone I knew in Key West. When she went back to her counter, I only sat and stared out of the window at the yellow glow of the sunset and at the yellow glow of the streetlights. South Florida seemed like an eternity away.

A few years ago I took my family to Key West for a short vacation just so they could see what it was that attracted me to that part of the world. When we turned the corner to drive through the downtown area, there were crowds spilling onto Duval Street from every bar and restaurant and souvenir shop. The sidewalks were jammed with Hawaiian shirts and Bermuda shorts and black Oxfords and Tilley hats in some kind of international bazaar that moved in Fujicolor waves from one corner to the next. I wanted my family to see Key West through my eyes but several thousand cruise ship tourists, smelling of night-blooming jasmine and stag-country aftershave, swarmed along Duval like sunburned Huns. I took a deep breath in the air-conditioned Taurus and drove down the street. Key West didn't seem the same.

There were crowds back in the mid-seventies, too, but we were mainly a loose community of regulars who moved from the Green Parrot Bar and Sub Shop to Captain Tony's to the Full Moon Saloon to Sloppy Joe's and sometimes all the way over to the Half-Shell Raw Bar down by the Standard docks where there was good Dixieland jazz on Sunday afternoons. Hemingway's Key West was just a backwater town for all of us, a town with a Bahamian beat and a Caribbean spirit where the shadows fell over the quiet streets and the palms sagged low in the heat, and where you could walk for blocks down Angela and not meet another soul. At night we were part of a moveable fiesta that ebbed and flowed through the downtown bars, sometimes continuing until sunrise at the all-night beer joints out on Stock Island.

I looked through the windshield from inside the Taurus, hoping some of the local fiesta might still be around. We drove all the way down past Sloppy Joe's to Mallory Square, but there were only cruise ship tourists swilling about and a cloud cover that would hide the sunset. I turned the car around and headed back up Duval where I turned left on Angela because I was curious whether the real Key West might still be hiding in the neighborhoods.

I shouldn't have worried. Angela, Southard, Fleming, Eaton and Caroline, and all the alleys and lanes in between, had settled into the sub-tropical twilight, oblivious to the chaos of Duval. We drove the streets slowly and backtracked a few times just so my family could look at a conch house again or another widow's walk, or just to see the palms that crowded a yard, draping lacy fronds over a picket fence. By the time darkness had settled in, Key West seemed alive and well again.

The following day we drove down Whitehead and parked across the street from the Green Parrot. I stood on the corner and stared at the building that looked old thirty-three years ago, and even back then I didn't know what held the roof in place, but there were hidden emotions sneaking through I hadn't expected. My family and I crossed the street and ducked through the double doors so I could show them the corner seat where I sat for all those months drinking beer. The seat and the bar looked just as beat up. I leaned on my elbows and thought about how beat up I felt when Janie left me and I spent my time wallowing in self-pity, drinking bottles of St. Pauli Girl, one after another, until the bartenders closed the place down.

When Janie packed her things and moved off of *Sun Flower* and away from Key West, it nearly killed me. I hadn't done anything wrong and when I bar-hopped along Duval in 1976, at least I could walk with my head up. There were issues that were never resolved, though, and as Jimmy Buffett finally admitted for me, it was my own damned fault. The corner seat in the Green Parrot helped me beat myself nearly to death over it.

I kept looking at the bar and at the junk on the walls and at the general state of grunge on everything. I had half a mind to go into the restroom just to see if the same rusty porcelain water trough was still there. I stayed in the corner seat, puzzled again over what had happened with Janie. Casey and Marielle never knew her, but my wife Kathleen did. She might have wondered on our visit how I felt about returning to the very spot that was the Badwater point in my life. I didn't want to talk about it. There is no sniveling allowed in the Green Parrot, anyway. I drank a glass of tonic water instead of a St. Pauli Girl and then we left the bar behind to drive over to Jose's Cantina where a smiling waiter took our orders for Cuban pork chops, black beans and yellow rice, plantains in sweet syrup, and a Cuban Mix for Marielle. While we waited for dinner to arrive, I asked for a cup of Cuban coffee. I took the first sip and remembered, for all the drinking and all the late nights and all the agony over a failed marriage, it was the Cuban coffee that woke me and shook me to the core each morning.

There were coffee kiosks on a lot of street corners back then. On my way to work it was nice to stop for a cup and hang around in the early morning light. The coffee was so good everyone stayed to have more. We talked of fishing the weed lines out along the Stream and about the Northers that blew down and made it so cold you could see your breath roiling like steam over the two-handed cups. On those cold mornings when the wind blew hard, we huddled in the lee of the kiosks and drank the good Bustelo and laughed about being at sea on days like that. I stood near the old fishermen and laughed right along and I felt, for a time at least, that I was part of their world and Janie had never happened.

I turned to watch the counter girl again wondering what it was about her that made me think of Key West. There is a freedom to her movements, a loose and happy attitude that made me smile, and maybe that's all. There were a lot of girls like her tagging along with the moveable fiesta. Sometimes I wonder where they went, but it doesn't matter anymore. There have been a lot of happy years since then. Even though they say you can't ever go back, at least you can enjoy a few moments of reflection while sitting in a shopping mall Starbucks. The coffee is always good, even if it isn't Bustelo.

CHAPTER FORTY-SIX

the agonies of my father...

I found a seat today at a low counter that faces the north side of the Starbucks. There is a poinsettia on my left fairly exploding from a plastic pot covered with green foil. It reminds you of all those rosy cheeks you see on Christmas mornings. There is another poinsettia on my right that doesn't look so good. The green leaves are dry and shriveled and some of the red petals have curled and fallen and it makes you wonder what the difference is, why one poinsettia is the picture of health and the other is withered and wilted and dying.

There's a man sitting outside at a table beyond my window who is reading an article in the evening paper. The leader states in bold print that elderly women who smoke feel worse than those who don't smoke. The man reading the article is puffing away on a cigarette. I watched as he reached for the ashtray on his right and flicked the cigarette lightly to knock the ash from the end. He never took his eyes from the paper. Maybe he thinks because the article is about old women it has no relevance to him, but his skin is lined and his face has taken on the color of canning wax. With all the smoke around his head, you wonder what his lungs look like. I'm glad there's a window between us.

If you watch someone smoking, it's hard not to notice things like the two poinsettias on the counter to my left and to my right. The one is big and full of life and its color is just like the mortars that fill the sky on New Year's Eve. The other poinsettia is like the man outside with the ashen face, the one who is smoking like a chimney while reading about the tired and shrunken old ladies who don't feel so well.

Cigarettes killed my father. On the day of his funeral, my brother and my three sisters sat around the dining room table crying and laughing and trying to cope with the emotions of love and hate and fear, and mostly the confusion of living with a man like my dad. While they were in the throes of all this agony, they puffed on cigarettes one after another and the room was filled with smoke. I decided not to get involved in the talk and so I sat alone on the couch and listened. Sometimes I laughed with them and sometimes I became enraged at them, but I remained silent and watched the layers of smoke drift slowly up and down in the faint currents of air that ghosted through the living room.

I remembered being a kid and watching my dad smoke while he read the evening paper. If I remained still enough, the smoke formed layers that drifted up and down in the faint air currents. When it was finally calm, I jumped from the floor and flailed my arms to set the smoke swirling and I would lie on my back on the floor again to watch the patterns. I was innocent of any knowledge of death and so was my father who smoked two or three packs of unfiltered Chesterfields each day. The coughing and the wheezing would only come in later years. My innocent father enjoyed his Chesterfields like he enjoyed his Ancient Age and I enjoyed playing in the layers of smoke that floated in our living rooms. I kept waving my arms and swirling the smoke, thinking all the time he would notice me.

When the doctors finally told my dad he suffered from emphysema, he wouldn't acknowledge his condition had been caused by cigarettes. He didn't want to admit he had been wrong all of his life, but the coughing became violent and he couldn't catch his breath. He quit cigarettes when it became nearly impossible for him to smoke. The quitting came too late. He was only sixty-seven when he died, alone in a motel room in Quincy, California, because he insisted on driving up to the Sierras even after the doctors told him to avoid the altitude.

The Chesterfields destroyed my dad. While I watched the smoke in the living room and listened to the rest of my family dealing with their grief on the day of his funeral, I simply didn't understand. I got up from the couch and walked through the layers of smoke to open the front door. I stepped outside onto the porch and breathed deeply while the tears made their way down my face. I felt in my heart the agonies of my father when he was alive, gasping for breath even when the air was clean and pure.

My family quit smoking soon after my dad's funeral. When I think of my sisters and my brother, I'm glad the madness of the cigarettes is no longer part of their lives. The man who was smoking outside the Starbucks window has gone away. He left behind the paper with the article about the old ladies who smoke. He took his Marlboros, though, and is probably smoking in his car. I'm looking again at the crimson leaves of the poinsettia on the counter. The one to my left. The healthy one.

⁄) CHAPTER FORTY-SEVEN ⁄)

sets on rewind...

A high overcast today gave way to a blue spring sky with cloud formations that looked like the canyon country over near Moab, but the curtain came down on the afternoon when a thick fog rolled in from the sea. The sun has gone down and the fog has lingered and now I'm sitting in the bright lights of the coffeehouse looking through the windows at the gray that swirls across the parking lot. Beneath the light standards the droplets mill about and form eddies and spirals and blow thin and then thick, and the light is first bright and then diffused. If you stare long enough, you get the sensation it's not the fog riding currents of wind but, rather, the parking lot and the storefronts and the light standards that are traveling through the mist. It's like looking at a time machine.

A few minutes ago an old man was standing outside with a cigarette in his mouth. When he exhaled, the smoke swirled about his head like the fog in the parking lot. The smoke and the fog and his hair were the same color and they mixed together briefly and made it look like the man had his head in a cloud all his own. He threw the cigarette on the sidewalk and stomped on it with his shoe and then he turned and pulled his frayed jacket together and hunched his shoulders against the wind and fog and then disappeared into the mist in the parking lot. It's dark where he once stood. You sit here in the bright lights and wonder what brought the old man to this point and where he might have gone. Maybe he had no place to go but into the night alone. You wonder how much a man like that would give to be walking into a real time machine, one that swirls in the fog and diffuses the light in the parking lot and sets on rewind an old man's life.

My son Casey, who is now twenty-three, doesn't much care for coffee. While he sat in Starbucks with me the other day, we started talking about what the future might hold. He is a recent graduate of the University of California up in Irvine and I could see a little apprehension in his eyes as he spoke of what the coming years might bring. I saw the same look in my dad's eyes all of his adult life, but with Casey I hadn't noticed it before this. Maybe it's just a sign of growing up. I worry just the same.

Our toilet stopped up one day when Casey was just a toddler. While I stood over the bowl and tried to get it to clear, my little boy pointed to the swirling water and said, "Bubble, Daddy. Bubble." I nodded at him but the toilet wasn't clearing no matter how I used the plunger or the snake. I was running out of patience. Little Casey just pointed to the water and said again, "Bubble, Daddy. Bubble…"

I started to get angry then and I blurted out, "I know those are bubbles. I'm trying to unclog it. Now go find your mommy."

Casey didn't leave, though, and I kept plunging the toilet and twisting the snake and making myself even angrier when Casey pointed again to the water in the bowl and with a serious little-man voice he said, "Bubble…Bubble," and that's when I lost all my patience.

"I know those are bubbles, God damn it, now get out of here and let me try to fix the toilet."

Tears welled up in his soft brown eyes and he looked in the bowl and he said weakly, "Bubble, Daddy…" and then he ran crying to his mom in the kitchen.

I kept swearing to myself and trying to clear the toilet until I finally got so frustrated I gave up and called for the plumber who was going to charge me an arm and a leg for a job I failed to do. When the man arrived, though, he couldn't free the toilet either. He unbolted the whole thing from the bathroom floor and found wedged inside one of Casey's bottles of milk. Wait a minute. Bubble? Bottle? Bubble of milk in the toilet? It dawned on me then that little Casey with his fifteen-month-old vocabulary was just trying to tell me his "bubble" was in the toilet and that's what stopped it up. I stared blankly at the mangled plastic and I looked at the plumber who just stood there and then I walked into the kitchen where Casey clutched at his mother's legs. I picked him up and held him close and I tried to apologize for being so angry and stupid, but there were some tears of my own caught in my throat and some of them made their way down my cheeks. I wished more than anything there was a way to start the morning over again.

Casey says he doesn't remember that day. I do and when I watch him now, I am still aware of all of those scars I left behind. He is a happy kid and far brighter than I am, but when I looked in his still-so-soft eyes while we talked the other day, I noticed again those same faraway looks I remember seeing in my dad. Casey has a future to sort out and you can tell it worries him. Sitting here in the coffeehouse with the fog outside swirling about and the yellow lights of the parking lot going dim and then bright again, it makes me realize I have to back away and let him find his way.

During my time in Key West when I spent my evenings at the Green Parrot, I got tired once of the noise and laughter and non-stop partying. I wandered by myself over to the Pier House for dinner. When I got there a good crowd of people waited for tables and even my friend Cindi who worked there as a receptionist couldn't get me a seat. I left and walked outside where another crowd had gathered on the little beach below the restaurant to listen to the Bahamian calypso band. I stood in the background watching while the tourists in their blazing tropical shirts danced stiffly in the sand. There were a lot of mai tai's and rum punches and bottles of beer on the tables and a lot of laughing and stumbling for the tourists. I left after a few minutes and walked into the bar for a St. Pauli Girl.

The Chart Room was completely empty except for the guy pouring drinks. I took a seat on a stool at the bar and ordered a bottle of beer. I hadn't been in the Chart Room in a long time and I sat there looking about at all the nautical décor hanging on the walls. I remember thinking how phony all of it seemed when a tall man wearing glasses and a faded Hawaiian shirt wandered in and sat a few seats away. My barroom rapport is worthless so I didn't speak to him. He knew the bartender and so there were some chuckles and some quiet comments and then he turned to me and introduced himself. I smiled and told him I was a yacht delivery captain between engagements, marital and professional, and he laughed at the joke and asked if I were a diver as well. We chatted for a long time about his diving business and how he had been searching for the wreck of the Spanish Galleon *Atocha*, and then he offered me a job at minimum wage. He couldn't afford more, he said, but he was giving his divers some shares in the treasure if it was ever found. I kept thinking about the endless hours of diving, some of it beyond the saturation point I was sure. When I left the Chart Room, I thanked the guy and then I declined his offer. Minimum wage just wasn't enough it seemed, especially when being asked to risk your life for a treasure that might never be found.

After I slipped out of the bar to head back to the Green Parrot, the wind felt good on my face. It ruffled my shirt and rustled the fronds of the palms that lined the street. I kept thinking while I walked in the dark about Mel Fisher and his treasure hunt. I knew he had already lost a son to a capsize accident down near the Marquesas Keys, but when

I wandered out of the night air and into the Green Parrot, Matt and Franco and Big Jim and Drew and the others were drinking and partying and trying to save my corner seat. I ordered a sub sandwich and a draft beer for dinner and when I sat in the bar stool, I was swept away one more time by the laughter and the music and by the Green Parrot regulars who danced and yelled and screamed in the heat trapped in the little wooden bar at the corner of Southard and Whitehead.

Up the freeway from where I live now there is the North County Faire, a mall where you can find restaurants and shoe stores and candy shops and where some years ago there was a road show at one of the jewelry stores. Out of curiosity my wife and I drove up to see it. There were anxieties creeping in for me and moments of regret I couldn't shake even when laughing about it with my wife. We walked into the store where the salesmen circled several display cases like hammerheads. We ignored them and looked silently at the emeralds and the gold chains and the bars of silver and pieces of eight and at the mounds of gold coins, all taken from the wreck of the *Atocha* by Mel Fisher and his crew of divers who worked for minimum wage.

One of the salesmen traveling with the show came near and in a moment of frustration I told him about Key West and the Chart Room and the bottles of St. Pauli Girl and about talking to Mel Fisher. When I told him about turning down the two-dollar-an-hour offer and walking out into that cool Key West night, he looked at me with a stunned expression on his face. He just shook his head. I didn't stay very long after that. When we left to return to our car, there were moments where all I could think about was my teacher's salary and the huge debt we had incurred when we bought our house and about Casey and Marielle and all of their college expenses, and then there were the fresh images of all those gold and silver coins and the ingots and the emeralds and those incredible chains. I hugged Kathleen and opened her car door. I don't think she ever knew how desperate I felt.

When I see the look in Casey's eyes that remind me of my father, at least I know he's at an age where the future is a greater concern for him than money or sunken treasure. He has wisdom beyond his years and it isn't likely he will wander through life searching randomly like his father and his grandfather before him. I wish I could gather Casey up and squeeze him close and try to fix everything by waving a magic wand and going back to the days before I ever lost my boat, before I ever got involved with the dope runners, before I ever turned down a chance at the treasure of the *Atocha,* or maybe just before I ever swore at him for his bubble in the toilet, but the droplets milling around outside that keep diffusing those yellow lights are only just fog. There is no time machine.

CHAPTER FORTY-EIGHT

crystal mosaics, lightly smeared...

It's another late night in the Starbucks and I'm alone near the front windows where I have a good view of the empty parking lot outside. An older man and woman near me were laughing so loud I thought they had been drinking, but when they left, they smiled at me and asked how the writing was going. I was caught off guard by their kindness. I only smiled and said it had been a struggle. They were a nice couple and I'm sorry I didn't listen to what made them laugh. The coffeehouse seems hollow without them. I'm the last customer of the night and without the two old people here, there's nothing left but the yellow legal pads on my table and the trash spilling from the cans outside.

A rain fell so fine this morning the drops didn't really fall but settled to the ground like heavy fog. When the sun broke through the mist, a towering rainbow arched beyond the hills to the west. It arched so high there was no end to the rainbow and it seemed like the colors faded into the gray of eternity. The rainbow only stayed for a moment and then the clouds shifted and hid the sun and wiped the colors from the slate of the sky like someone cleaning fingerprints off the windows of the coffeehouse.

A few years ago in Durango where I first began to write, I drank coffee for long stretches in the Steaming Bean and watched the customers come and go. I also spent a good part of the time staring through the fingerprints that covered the windows. They looked at times like a thousand silver-gray clouds. No one ever came to clean them off and after a while they only seemed to be part of the decor. You could see the fingerprints of the children smudged down low while those of the adults swept upward, disappearing into the dust near the shadows of the eaves. It made the windows look like crystal mosaics, lightly smeared, of the sidewalk people and the people of the Steaming Bean.

I sat looking through the fingerprints one day when a middle-aged woman walked into the coffeehouse. She wore a straw cowboy hat with a lot of feathers stuck into the band at odd angles and carried with her a soiled and frayed bedroll she dropped carelessly on the floor by the line of bar stools. The girl behind the counter gave the lady a free cup of coffee. The lady never smiled, only muttering a thank-you through teeth the color of butterscotch and road tar. I watched as the lady took her cup of coffee and walked briskly through the door, leaving her bedroll behind and mumbling nonsense to herself as she left. She went out with such purpose I thought she might just be going out for a cigarette. She didn't come back. Her bedroll sagged against the counter under the feet of a man who pushed it aside with the toe of his boot.

When I walked to the coffeehouse through the mist that morning, a raven in a tree squawked at me. Each time it squawked, I counted to four. It seems like an animal with so little intelligence would squawk in a random pattern in response to fear or to attract the attention of others and that the number of squawks would vary, but the raven in the tree only squawked in fours. When the mumbling lady in the cowboy hat walked by again, she didn't come in to retrieve her bedroll. There were four newspaper racks outside. She stopped in front of each of them to check for loose change. Four times she pushed the coin returns. Four times a loud squeak drifted into the Steaming Bean. I thought of the raven in the tree staring blankly at the world below.

There were no coins that day. The lady stared blankly at each of the machines and then she turned and disappeared up the street beyond the windows. I sat in the Steaming Bean and stared after her, like gaping at an empty parking space after your car has been stolen. I could only wonder what it must be like to blink at the world like a raven that squawks from a tree. Someone's heart is broken because the lady in the cowboy hat didn't get out of the Sixties alive. The only fingerprints she makes are smeared on the cold storefront windows of Durango.

I stopped for coffee at the Jackalope Café in Cave Creek, Arizona. When I pulled my truck into the parking lot, there were a dozen or more Harleys out in front including a half-dozen choppers. I looked at them briefly and when I went inside, I expected to see greasy leather and greasy hair and greasy Nazi helmets, but there were only doctors and lawyers and schoolteachers. The whole place was clean and civilized. I sat in the corner near an old couple and tried to write. It was a struggle again and when the Harley riders left, I only sat and listened to the throb of the bikes as they roared from the parking lot.

There is no other sound in motor sports like the big Harley twins. When the first of them started up, the plate glass window in the front of the café resonated and shook with the concussions. The old people sitting nearby looked with their mouths open at the doctors and lawyers who were leaving to continue their Sunday ride. I thought the window might shatter, but I wasn't sure the old couple had even heard the Harleys. The man tried to lift a spoonful of cornflakes while his hand shook like the window, and the milk from his spoon dribbled onto the table. I didn't think the old people would live very much longer. When I left to pay my bill, I saw that there were no fingerprints on the plate glass window. The Harleys had roared off toward Scottsdale leaving behind only the silence and the stillness of the high desert air. The shaking old man could only sit and stare. I wondered what kind of fingerprints the old man would leave behind.

If my dad had lived, he would have been about the same age as the man spilling milk in the Jackalope Café. When I drove out of the parking lot in my truck that day, I kept thinking about the old man and it made me wonder about age and about that point in time where the quality of your life has diminished so far that living itself becomes a burden. My father, even with his drinking, never had a trembling hand like the man I left behind in the café. I wonder how long he would have hung on if he couldn't even drink his morning coffee?

In all of our time together as a family, my mother held to certain standards that she simply wouldn't abandon. When supper was served, there were always table cloths and napkins and silverware and serving dishes and never did she ladle a vegetable or a main course from a pot on the stove to a dinner plate. For breakfast she made my father coffee every morning, pouring it into delicate cups and saucers. My father tolerated my mother's quirks, but when it came to his morning coffee, he didn't want to wait for it to cool. He dribbled the coffee from the cup into the saucer until it nearly overflowed. He then lifted the saucer to his lips and sipped the cooling coffee. Never once did I see him spill it. My mother always gave him a disgusted look, but there was a smile hidden in there somewhere. My dad would glance from behind his saucer and wink at me.

I can't imagine my father being so old that he could no longer tease my mom with his saucer of coffee. The man in the Jackalope Café who could hardly hold a spoon was far beyond that point in his life. At least he was around to try. My dad might have accepted the ravages of time, but while I drove down the Arizona two-lane toward the Interstate in the distance that morning, I felt somehow cheated that he couldn't give himself the chance. Now there are no more of his fingerprints.

The girl who works the counter here has told me the Starbucks will be closing in a few minutes. She asked me what my writing was all about. I told her I didn't know, that maybe it was just a mosaic. She only smiled at me and went about wiping the tables clean with a rag. I stood and walked over toward the windows in the front and held my coffee with two hands and stared once more through the fingerprints. I wanted snow flurries again and long, empty highways curving into nothingness in the distance. I wanted my dad to be staring out of the windows beside me, holding his cup in two hands with the coffee steaming in the cold air, but it's dark and still outside. The yellow sodium vapor lights in the parking lot make the place look like a carnival at closing time, not a two-lane highway in Little River.

After having been around for over a year to write, it's strange to find that on my last day here, I'm alone. When I walk outside into the night, I will leave my fingerprints on the glass door of the coffeehouse, but then the girl will wipe the surface clean with her rag. It will seem like my dad and I were never here. Maybe when I'm gone it would be nice if someone made a comment about the guy who sat writing in the corner by himself, the one who tried to see the horizon beyond the windows. Maybe it would be better for them to think of fingerprints and snow flurries and a fading rainbow from the morning rain, the one that searched for openings in the sky.

THE COVER

The map gracing the cover, a section of the Western Atlantic north and south of the Tropic of Cancer, was taken from an original cartographic work entitled *Insulae Americanae in Oceano Septentrionali cum Terris adiacentibus,* engraved by Pieter Schenk and Gerard Valk in Amsterdam, circa 1690. This particular map was chosen for its geographical significance relative to the stories in this text and is an accurate reflection of the Gulf of Mexico, the Bahamas, and the northernmost Caribbean islands as imagined by an eleven-year-old kid fishing in the Low Country of South Carolina.

Permission for the use of the map was granted by Barry L. Ruderman of Barry Lawrence Ruderman Antique Maps, Inc., of La Jolla, California (www.RareMaps.com). Mr. Ruderman was kind enough to share his extensive collection with us and provided a high resolution scan of the original engraving for the cover. In exchange for this amazing generosity, he asked only for a copy of this book. I owe him far more.

THE AUTHOR

Lew Decker is a retired middle school teacher who makes his home in San Diego, California. He has held more than fifteen positions in a variety of career paths ranging from cartographer to charter boat captain, including a stint in the U. S. Navy during the Vietnam era. A veteran blue-water yachtsman, Lew spent several years in the Caribbean sailing circuit and is the only person on his block who has been shipwrecked on a deserted island. Today he enjoys adult league baseball, fly fishing, road trips, and amateur radio (KJ6G), and is an avid follower of vintage car racing.

www.ingramcontent.com/pod-product-compliance
Lightning Source LLC
Chambersburg PA
CBHW022005090426
42741CB00007B/895

* 9 7 8 0 9 8 4 0 9 7 1 0 4 *